Social Empiricism

Social Empiricism

Miriam Solomon

A Bradford Book
The MIT Press
Cambridge, Massachusetts
London, England

First MIT Press paperback edition, 2007

This book was set in Sabon by Achorn Graphic Services, Inc.

Library of Congress Cataloging-in-Publication Data

Solomon, Miriam.

Social empiricism / Miriam Solomon.
p. cm.
"A Bradford book."
Includes bibliographical references.
ISBN 978-0-262-19461-7 (hc : alk. paper), 978-0-262-69352-3 (pb)

1. Science—Philosophy. 2. Knowledge, Sociology of. I. Title.
Q175.S652 2001
501—dc21

2001030623

To my father, Norman Solomon, and in memory of my mother,
Devora Solomon (1932–1998)

Contents

Preface

Social Empiricism took shape over a long period of time, and with the influence and support of a number of people and institutions. My former colleagues at the University of Cincinnati supported my return to philosophy of science, after a useful graduate school detour in the history of analytic philosophy. A Mellon Postdoctoral Fellowship at the University of Pennsylvania in 1990–91 gave me the time and the environment to learn about new developments in the cognitive sciences and in science studies. As a participant in the NEH Summer Institute, "Science as a Cultural Process" in summer 1991, I read widely in science studies and feminist science criticism, and had the opportunity to argue with authors. It was then that I saw the emptiness of the polarized debates between social constructivism and traditional empiricism, and looked for the emergence of some new, more complex ideas.

During the early 1990s, it became clear that full development of social empiricism would require several careful historical case studies. The first case study that I did—the history of continental drift and plate tectonics—was suggestive, but not enough. I chose to spend a year on an NEH Fellowship for University Teachers (1994–5) at the Dibner Institute (MIT), which made up the difference in my salary and gave me the opportunity to learn from other Fellows working in the history of science. During that year, in my office overlooking the Charles River and the Boston skyline, I began writing this book.

Temple University, my academic home since 1991, has supported my work with several summer grants and a semester of research and study leave (Fall 1996). *Social Empiricism* was both delayed and improved by several sideline projects in cognitive science, feminist philosophy of science and bioethics. The impending arrival of my daughter, Amira (born on April 6, 2000), gave me the incentive to finish the manuscript and submit it for publication.

I could not have written this book without the help of a number of colleagues, friends and graduate students, who read portions of the manuscript, answered factual questions, debated broad issues and fine points, and provided continuous moral support. In particular, I thank (in alphabetical order) Liz Anderson, Sid Axinn, Dick Burian, Stephen Downes, Chuck Dyke, Gary Ebbs, Moti Feingold, Samuel Freeman, Ronald Giere, Alvin Goldman, Gary Hardcastle, Jonathan Harwood, Gary Hatfield, Cliff Hooker, Mark Kaplan, Donna Keren, Philip Kitcher, Hilary Kornblith, Hugh Lacey, James Maffie, Mimi Marinucci, Nancy McHugh, Linda Weiner Morris, Ted Morris, Nick Pappas, Ellen Peel, Joan Richards, Alan Richardson, Bob Richardson, Marya Schechtman, Fred Schmitt, Paul Thagard, Jerry Vision, Joan Weiner and Alison Wylie. I have benefited from the generous comments of my audiences at talks at various universities and conferences. Referees of this book made useful criticisms, corrections and comments. My husband, John Clarke, cared about this project as if it was his own, and gave me both love and time.

The material in chapters 4–7 of this book is developed from previously published work. I am grateful to the publishers of the following articles for permission to include some ideas and short excerpts from them:

Solomon, M. (1994). "Social Empiricism." *Noûs* 28, no. 3: 325–343.

Solomon, M. (1994). "A More Social Epistemology." In Fred Schmitt (ed.), *Socializing Epistemology*, pp. 217–233. Lanham: Rowman and Littlefield.

Solomon, M. (1994). "Multivariate Models of Scientific Change." *PSA 1994*, vol. 2, pp. 287–297.

Solomon, M. (1995). "Legend Naturalism and Scientific Progress." *Studies in the History and Philosophy of Science* 26, no. 2: 205–218.

Solomon, M. (1998). "Happily Ever After with Consensus?" *Feneomenologia e societa* 21, no. 1: 58–65.

Solomon, M. (2001). "It *Isn't* the Thought That Counts." *Argumentation* 15, no. 1, pp. 67–75.

Solomon, M. (2001). "Consensus in Science." In Tian Yu Cao (ed.), *Philosophy of Science*, vol. 10 of *Proceedings of the Twentieth World Congress of Philosophy*. Bowling Green, Ky.: Philosophy Documentation Center, pp. 193–204.

Social Empiricism

1
Introduction

If you turn your back on Mishnory and walk away from it, you are still on the Mishnory road.
Ursula LeGuin, *The Left Hand of Darkness*

Two disputed claims have been at the core of discussions about scientific change for the last forty years. The first is that scientists reason *rationally;* the second is that science is *progressive.* For about thirty years, the controversies were cross-disciplinary, led by philosophers on one side and sociologists of scientific knowledge on the other. Discussions were polarized, with philosophers defending traditional Enlightenment ideas about rationality and progress while sociologists of scientific knowledge (often abbreviated SSK), who extrapolated some themes in Kuhn's work (1970 [1962]), espoused relativism and constructivism.[1] Historians and others in science studies were sometimes guided by philosophers, sometimes by sociologists of scientific knowledge and sometimes, in two-voiced narratives, by both.[2]

Recently—over the past ten years or so—the debate has become less polarized. On both sides, careful examination of case studies has led to new ideas about rationality and progress. Creative ideas have come from science studies disciplines in addition to philosophy and sociology of scientific knowledge, such as history of science, feminist science criticism, psychology of science and ethnographic studies. The result is not a compromise or a midpoint between two extreme positions; it is the beginning of a fresh understanding of scientific change.

Social Empiricism is inspired by and addressed to this new community of ideas. The goal is a systematic new epistemology of science. This new epistemology is not intended to be the last word on scientific change. Rather, it claims to move beyond piecemeal new insights and remnants of past unworkable positions, advancing the debate to the next stage.

This introduction begins with an analysis of the immediate post-Kuhnian debate between philosophers of science and sociologists of scientific knowledge. Then it surveys the more recent suggestions, showing how they reject the framework of the original debate. The introduction ends with an outline of central themes in this book, organized by chapter contents.

In my view, philosophers of science and sociologists of scientific knowledge had *common* assumptions about the nature of scientific rationality and scientific progress. They disagreed about whether scientists reason rationally and whether scientific change is progressive, but they had shared standards of evaluation. For all, the framework of debate was Enlightenment epistemology. Descriptions of this framework and the opposed positions of traditional philosophers of science (e.g., Hempel, Lakatos, Laudan, Glymour and some early work in naturalistic philosophy of science such as Giere 1988, Thagard 1988) and sociologists of scientific knowledge (e.g., Barnes, Bloor, Collins, Pinch, Woolgar) give evidence for this interpretation.

Some of the shared assumptions about the nature of rationality are as follows:

• *Individualism* Rational thinking is thought of as an ability attributable to individual human beings and a property of their decisions. Scientific rationality, which is one kind of rationality,[3] is thought of as a property attributable to the decisions of individual scientists and typically exercised by those scientists (e.g., Galileo, Newton, Darwin, Einstein) historically credited with progressive scientific change. The term "rational" is not applied to groups or communities, except derivatively (a rational community is composed of rational individuals). The debate between traditional philosophers of science and sociologists of scientific knowledge concerns how *frequently* individual scientists reason rationally; the former think irrationality is unusual, while the latter think that rational reasoning is unusual. For example, Larry Laudan's well-known "arationality assumption" (see, e.g., Laudan 1987, p. 202) recommends that nonrational causes of belief be looked for only in the (according to him) rare case when there are insufficient rational reasons for a scientist's decisions. Naturalistic philosophers (e.g., Giere 1988) continued to argue for individualism by claiming that it is cognitive (i.e., individual psychological) factors, rather than social ones, that produce scientific change. Sociologists of scientific knowledge, on the other hand, have inherited Barnes and Bloor's (1982) "equivalence postulate," which claims to find that all beliefs—plausible and implausible, apparently well justified and not, true and false—have similar causes in human interests. All look at the causes of individual beliefs and draw conclusions about individual rationality.

• *Objectivity (including "cold" cognition)* It is assumed that scientific rationality is thinking that is free of motivational and ideological bias. During the 1980s, some philosophers of science (e.g., Giere 1988, Thagard 1989) put a naturalistic gloss on this assumption, claiming that scientific rationality is "cold cognitive processing," i.e., information processing motivated only by the desire for "truth" or other purely cognitive ends (e.g., good models, predictive success, explanatory success). They looked to the newly flourishing science of cognitive psychology to find out what the processes are. Sociologists of scientific knowledge countered that all scientific decision making is motivated by personal or social goals (i.e., is "hot" cognitive): they find that factors such as self-interest, the influence of authority, peer pressure, pride, conservatism, ideology, enthusiasm for new technology and funding practices determine individual scientists' decisions. Because they also identify rationality with "cold cognition" and consequent freedom from motivational and ideological bias, they conclude that scientific decision making is not rational. Latour's ten year moratorium on cognitive studies of science (1987, p. 247) was an attempt to focus attention on the importance of other causes of belief.

• *Method* It is assumed that scientific rationality requires following explicit or implicit rules of reasoning—such as probabilistic reasoning, rules of confirmation and disconfirmation, measurements of problem solving ability or explanatory power, and heuristics such as the principle of parsimony. Traditional philosophers of science (e.g., Glymour, Laudan) still aim to produce such rules, while sociologists of scientific knowledge argue (following Kuhn and Feyerabend) that the project is hopeless because successful scientists do not follow rules.

- *Generality* It is assumed that all scientifically rational work has in common the *same* scientific method. That is, scientific method is a *general* property of rational scientific decision making. The generality assumption lies behind the usual reading of Feyerabend's (1975, p. 23) provocative statement that "Anything goes": Feyerabend finds cases where suggested rules of reasoning fail, and then declares that there is no scientific method. This reading of Feyerabend's position[4] was taken up by SSK. Traditional naturalistic epistemologists, on the other hand, continued with their positive elaboration of the generality assumption, e.g., when Kitcher (1993, p. 10) devotes himself to "the task of recognizing the general features of the scientific enterprise."

There is also a trend among philosophers, encouraged by the writing of senior philosophers of science such as Hacking (1983) and Fine (1986), to eschew questions about scientific method and focus instead on ontological questions. For example, Hacking writes, "The 'rationality' studied by philosophers of science holds as little charm for me as it does for Feyerabend. Reality is more fun . . ." (1983, p. 16). Their reasons for this preference for the SSK reaction to Feyerabend are not obvious. While there are good reasons to abandon the hope for a general, algorithmic scientific method, there is every reason to expect the same kinds of insights for epistemological questions as Hacking and Fine expect for ontological questions, i.e., piecemeal, domain specific and historically contingent.

- *All-or-nothing* The traditional assumption about rationality is that it is an all-or-nothing property: one failure to exercise proper scientific method jeopardizes the rationality of the entire decision process of which it is a part. If the causes of a decision are 90% "scientific" and 10% "social," the jury on rationality is still out. Rationality is not measured in degrees. The insistence of the sociologist of scientific knowledge that *all* scientific change is social is, really, a claim that social (= non-rational) factors substantially influence all decision making and, therefore (assuming the traditional all-or-nothing view about rationality), that there is no scientific rationality. The traditional philosophical defense of rationality is the claim that social influences on decision making rarely or never affect the outcome (e.g., Laudan 1997, Thagard 1989). I call this the "Ivory Soap" model of scientific rationality, because the claim is that reasoning is pure enough to be considered epistemologically pure (99.44 percent purity is advertised as pure enough for bathroom soap).

Some shared assumptions about the nature of scientific progress are as follows:

• *Truth* The most widespread traditional view is that truth is the goal of science, and that science progresses by accumulating truths (or, at least, partial or approximate truths, or representations, or models resembling the world). Sociologists of scientific knowledge who find that there is scientific change without increased truth are skeptical of any claim that there is cumulative progress in science. (To avoid a pointless equivocation about the word "truth" here, I do not use it for beliefs which are socially constructed in a manner that is epistemically arbitrary.)

• *Objective* One way in which the progress of science can be objective is if it consists of the accumulation of truths. Such truths are not socially constructed or negotiated, nor are they relative to the theory under consideration. Other measures of progress (e.g., predictive success, explanatory success) can also be objective, so long as they do not result in judgments that are relative to person, social group or theory under consideration. The minority of traditional philosophers who are not realists (e.g., Laudan, Van Fraassen, even Kuhn on some readings) typically claim this kind of measure of objective progress. Sociologists of scientific knowledge are skeptical of *any* measures of objective progress. (Note that the term "objective" is also used in a related sense—interchangeably with "rational" and usually meaning "cold cognitive"—when describing the methods of science, above.)

• *Consensus* The traditional view about scientific progress is that scientific change is marked by consensus; significant change happens when one theory, or a set of competing theories, is replaced by *one* theory. Traditional philosophers talk of *rational* causes of consensus (e.g., overwhelming evidence for one theory) while sociologists of scientific knowledge talk of the *social and political* causes of consensus (e.g., the effect of the

Restoration on the acceptance of experimental methods in seventeenth century England).

• *Linguistic* The linguistic products of scientific activity are: explicitly stated explanations (which may be set out in traditional deductive form), propositions that claim to correspond to the world, derived predictions that claim to match observation reports. Traditional philosophers of science (e.g., Hempel, Laudan, Kitcher, Van Fraassen[5]) measure scientific progress in terms of these linguistic products. Sociologists of scientific knowledge argue that these products are not the objective achievements they appear to be, and go on to be skeptical about scientific progress altogether.

• *"Pure" science* Traditional philosophers of science claim that there is a clear demarcation between "pure" science and "applied" science (technology). There can be progress in one enterprise without progress in the other. Typically, "pure" and "applied" science are distinguished by having different goals of inquiry (e.g., truth versus building bridges). Sociologists of scientific knowledge deny that this clear demarcation is possible, and give up *any* distinction between science and technology. Latour (1987) coined the term "technoscience" to signify this.

Thus, shared assumptions about the nature of rationality and progress led to mirror image views of the nature of scientific change. Traditional philosophers of science and sociologists of scientific knowledge are closer to one another than they often believe. Their relationship is like that of authoritarian parent to rebellious child. The child (SSK) reacts against the parent (traditional philosophy of science) without changing the framework assumed by the debate.

A good deal of this post-Kuhnian exchange continues. Indeed, Pickering complains that "much of the work currently being done in science studies has remained stuck in the place where SSK left it in the early 1980s" (1995, p. 27). Since the late 1980s, however, new ideas that reject Enlightenment assumptions about the nature of rationality and progress are being widely produced. In fact, almost every assumption mentioned above has been jettisoned by at least one philosopher, historian, sociologist, feminist critic or anthropologist, and most by more than one. A quick survey gives examples of these developments.

First, on the nature of rationality:

• *Individualism* Social epistemology—a delayed response to both Kuhn and Wittgenstein (see Solomon 1996)—rejects the assumption of individualism. Philosophers of science such as Giere (1988), Hull (1988), Kitcher (1993), Solomon (1992) and Thagard (1993) have assessed scientific rationality during times of dissent from a social perspective. Instead of assessing the rationality of individual reasons and causes of belief, they assess the resultant distribution of cognitive effort by considering whether competing theories are each getting a fair share of investigation. Feminist critics, such as Haraway (1991), Harding (1991), Longino (1990) and Nelson (1990), have urged a more thoroughgoing social assessment of rationality. Longino in particular has specified four social requirements for rational scientific decision making (see 1990, chapter 4).

• *Objectivity* Giving up the assumption of individualism helps make possible the rejection of the assumption of objectivity in reasoning. This is because what matters for science is not the individual causes of belief (which may have motivational, ideological or even cognitive bias) but the resultant state of the scientific community, especially the distribution of research effort. Indeed, Kitcher (1993, chapter 8) acknowledges the role of decisions motivated by the desire for credit in rational scientific change. As he writes (p. 305), "particular kinds of social arrangements make good epistemic use of the grubbiest motives." I argued (1992) on similar grounds that "cold" cognitive processes are ipso facto no more "rational" than "hot" cognitive or even non-cognitive processes. Some feminist critics go further here, redefining the term "objectivity" (for clarity, Haraway and Harding call it "strong objectivity") in terms of particular political points of view, or specified social conditions that are especially conducive to a wide-ranging distribution of research effort.

• *Method* Kuhn (1977) proposed a non-algorithmic account of scientific method. Kuhn explicitly rejects the assumption of both traditional philosophers of science and sociologists of scientific knowledge that if there is a scientific method, it will be precise enough that there can be agreement on the promising direction(s) of research to take when following the method. Kuhn thinks that there can be room for disagreement among experts, but not so much disagreement that decisions are arbitrary. This positive view of Kuhn's has been reiterated recently, at least in general terms. For example, Pickering (1995, p. 32) writes, "On my analysis of practice, it is far from the case in science that 'anything goes'." Surprisingly, no one has developed Kuhn's account of objective but non-algorithmic theory choice.

• *Generality* Naturalistic philosophers (e.g., Goldman 1992, Kitcher 1993, Laudan 1984 and Boyd 1985) often claim that successful methods change as domains change or as science progresses. In other words, they claim that there is not a universal, general scientific method.[6] This philosophical position has been reinforced by Wittgensteinian arguments and by the recent interest in disunity of science claims (e.g., Dupre 1993). Actual studies of locally successful methods are rare, but have been produced recently. For example, Darden (1991) and Bechtel and Richardson (1993) explore and assess heuristics and reasoning strategies in the history of genetics, and Wylie (2000) explores a variety of heuristics in archeological reasoning.

• *All-or-nothing* Longino (1990, p. 76) claims that "Objectivity . . . turns out to be a matter of degree." Specifically, she acknowledges that her four conditions for objectivity—recognized avenues for criticism, shared standards (including, for science, empirical adequacy), community response and equality of intellectual authority—are typically imperfectly or only partly realized. The upshot is that "transformative criticism," which is at the heart of objective practices, comes in degrees of depth and range. For Longino, perfect objectivity is an ideal, and actual practices that do not attain this are still valuable for their partial objectivity. Wylie and Nelson (1998) also argue that objectivity comes in degrees.

Next, new ideas about the nature of scientific progress have come from the pragmatic tradition in philosophy, actor-network theory in sociology of science, and feminist critics. In general, there has been a shift from focus on theory (and thus on linguistic representation, truth, etc.) to a focus on experimental practice (where the emphasis is on what "works," where "works" is, unfortunately, never specified).

• *Truth* Hacking (1983) argues that knowledge about the existence of entities can be robust (firm, dependable) when we can successfully manipulate them. He remains skeptical about truth or even partial truth of theories. Increase in the knowledge of existence of entities, and how to interact with them, constitutes, for Hacking, progress in experimental science. This is "know how," which is not representational knowledge. Sociologists of science and feminist critics have expressed insights similar to Hacking's, in a more democratic voice: entities become known through their actions and resistances to our manipulations as much as through our successful manipulations of them. Latour (1987) and Callon (1986) use the semiotic terminology of "actants" in nature, Pickering (1995) writes of "material agency" and the "resistance of nature" to which we must "accommodate," Galison (1997) finds coordinated action in "trading zones" and Haraway (1991) celebrates the ability of nature to surprise us. For each of these writers, we need to adapt to nature (which they construe in various ways), and sometimes also to coordinate our interactions with one another, to get empirical success. Success, then, is not constructed by social interactions alone.

• *Objective* Apart from the claim that some practices "work," where "work" is not further specified, nothing is said. Even Kuhn's idea (1962) that there is "progress through revolutions" has not been taken up in any new way. As recently as Cartwright (1999) and Psillos (1999) the idea of "empirical success" is taken as unproblematic, and used as a proxy for "objective success," even though no account is given of it.

• *Consensus* Feyerabend's view, echoing Mill, that consensus is both undesirable and avoidable in scientific inquiry, has generally been ignored. Philosophers and sociologists of science typically still treat consensus as the goal of inquiry and as a natural resolution of dissent. Recently, a few feminist critics (notably Haraway 1991 and Longino 1990) have argued that pluralism, with continued dissent, is not an intrinsically unsatisfactory state for scientific knowledge. Rather, continued consideration of several theories acknowledges the partial virtues of all theories. For example, Haraway writes of "the joining of partial views and halting voices into a collective subject position" (1991, p. 196). Dupre's (1993) pluralism and Giere's (1999) "pluralistic perspectival realism" are influenced by Haraway's and Longino's work, especially.

• *Linguistic* Scientific success is not understood as a linguistic product (as a theoretical claim, explanation, explicit prediction) or even as a representation (including the pictorial models suggested by Giere 1988 and Nersessian 1992) but as a successful *practice* (see, for example, work of Galison 1997, Hacking 1983 and Pickering 1995). Of course, theoretical claims and models may play a role in *producing* the successful practice.

- *"Pure" science* There has been no recent change in this discussion.

Social Empiricism takes all these developments further. It has two parts, corresponding to the "social" and the "empiricism" parts of the thesis. Chapters 2 and 3 give a new empiricist account of scientific progress. Most of the book (chapters 4 through 8) develops and applies a new, social account of scientific rationality. Throughout, case studies, primarily in late nineteenth and twentieth century natural science, are used to develop the ideas.

Chapter 2, "Empirical Success," begins an answer to the neglected question, "what is the nature of empirical success?" and the related neglected question, "what is it for some practices to 'work'?" Chapter 2 gathers a range of examples of scientific success, distinguishes empirical success from other kinds of scientific success, and then discusses the robustness, significance and varieties of empirical success. Empirical success can be predictive success, experimental success, observational success, some kinds of technological success and some kinds of explanatory success; in each of these categories the empirical success can be more or less significant. Qualitative assessments of empirical success (such as robustness and significance) suffice for the purposes of this book. Some kinds of technological success do not come with significant empirical success; this point is the basis for a new understanding of the distinction between science and technology. The same is true for some kinds of explanation. Thus the contribution of types of explanatory success to empirical success is also reconsidered.

Truth is the most commonly stated goal of science, especially by scientists. Yet, frequently (as Laudan and others have pointed out), empirically successful scientific theories are substantially false in that most of the entities postulated by them do not exist and/or most of the claims of the theory are incorrect. It is tempting to conclude from this that the only genuinely attainable goal of science is empirical success. Chapter 3, "Whig Realism," argues that truth can still be a genuinely attainable goal of science. Whig realism states that there is typically *something* true about empirically successful theories, although such theories may not be literally true, partially or approximately true, or even good representations. Truth in theories is known, at best, in hindsight. (Empirical success, on the other hand, can be known during early consideration of theories.) Whig realism, unlike most forms of realism, is consistent with pluralism. Moreover, unlike most forms of both realism and antirealism, it has straightforward and reasonable methodological consequences.

In chapter 4 I introduce the terminology of *decision vectors,* which is used for discussions of scientific rationality in chapters 5 through 8. This terminology is epistemically neutral so that I can describe scientific decision making and scientific change without using such terms as "biasing factors," "external factors," "social factors" or "psychological factors," since these terms already come with negative normative associations. A variety of decision vectors, discovered in the various science studies disciplines, are described, in preparation for the work of later chapters. A distinction between empirical and non-empirical decision vectors is discerned, and this distinction will have epistemic importance in later chapters.

Chapters 5 and 6 explore a range of cases of dissent and consensus in science. Decision vectors influence all decisions—during dissent, during consensus formation, and during the dissolution of consensus—to the same extent, in all areas of scientific inquiry. Consensus can take place without general or central causal processes. Distributed models of decision making have recently been described in the artificial intelligence community as well as, more qualitatively, in the actor-network theories of Callon and Latour. These models are useful for understanding the cases of consensus that I describe.

Examples in chapters 2, 3 and 5 show that scientific success is not dependent on consensus, and examples in chapter 6 show that consensus is not always progressive for science. The move from dissent to consensus, therefore, is not in itself of much epistemic significance. The pursuit of empirical success and of truth can be consistent with both dissent and consensus.

Chapters 5 and 6 begin construction of a social epistemology for science, which is developed fully in chapter 7. It is a *more* social epistemology than those currently available. The historical case studies suggest that scientific rationality is socially emergent and not dependent on such conditions as individual clear thinking, rational decision making or reasonable inferences. Instead, it is a matter of having a particular distribution of decision vectors across the scientific community. Scientific communities are not merely the locus of distributed expert knowledge and a resource for criticism, but the site of *distributed decision making.*

To attest to the fact that the distinction between dissent and consensus is not one of much epistemic significance, and that, in particular, consensus is not a superior or a distinctive epistemic state, a normative account of scientific decision making—which I call *social empiricism*—is developed for dissent and then consensus is treated as a special case of dissent (where the amount of dissent approaches zero). *Social empiricism* is then applied to some new cases, and shown to yield useful normative recommendations. In particular, social empiricism is critical of the epistemic goals of the new institution of "consensus conferences" in the medical sciences.

Chapter 8, finally, reflects on the idea of "epistemic fairness" entailed by social empiricism. There is a parallel between epistemic fairness and new conceptions of justice, fairness and democracy in feminist political philosophy (e.g., Young 1990). This parallel reinforces my suggestions about the kinds of normative recommendations that social empiricism can realistically offer. Chapter 8 also considers the relation between social empiricism and the various feminist critiques of scientific inquiry. In my view, feminist critiques of science generally offer the deepest challenge to traditional views about scientific change. Social empiricism has much in common with them. Notably, both have a serious commitment to a naturalized approach to epistemology. I discuss several particular similarities and one important difference between feminist critiques and social empiricism.

Traditional epistemologies, from the time of Plato and Aristotle, and through the contributions of Descartes, Bacon, Newton, Mill and more lately Hempel, Laudan and others, have produced rules and heuristics for individual scientists. Social empiricism, while acknowledging the utility of *some* of these individual guidelines, develops rules and heuristics that are socially applicable. This means that the traditional focus on methods and heuristics to be individually applied by all working scientists is rejected. Instead, the normative emphasis is on science funding, administration and policy.

The case studies chosen throughout this book come mostly from late nineteenth and twentieth century science, and cover a range of sciences, from the physical sciences to the life sciences to the empirical social sciences. I have three reasons for these choices. First, I want to forestall any criticisms of the kind "what you say is applicable only to physics" or "what you say is applicable only to the 'softer' sciences" or "geology is an unusual case that does not generalize." Secondly, the scientific enterprise has changed over its long history, and it is likely that its epistemic character has also changed, especially with and since the Scientific Revolution of the seventeenth century. While it is plausible that my conclusions apply to all modern science (i.e., post-Scientific Revolution), I am being cautious and restricting my conclusions to a shorter historical period, i.e., recent and contemporary science. Finally, my purposes are practical rather than historical or abstractly philosophical. My goal is to positively affect scientific decision making through practical social recommendations. For this reason, I focus my investigations on the past 100 or so years, including contemporary as well as recent historical cases, so that there will be no questions about relevance to contemporary issues.

Quine famously regarded science as an extension of common sense (1966, p. 229). For him, and many others, answers to epistemological and ontological questions about science can come from reflection on ordinary practices. Even Pickering (1995, p. 6) stresses the continuities of science with the ordinary: "Much of everyday life . . . has this character of coping with material agency, agency that comes at us from outside the human realm and that cannot be reduced to anything within that realm." Certainly, general epistemology and philosophy of science have positively influenced one another. Yet, philosophy of science is not epistemology. Science may have developed from ordinary practices, but it has developed distinctive practices, particular goals and unique forms of social organization. Social empiricism is an epistemology for science.

2

Empirical Success

We are not in charge of the world.
Donna Haraway, *Simians, Cyborgs, and Women*

1 Scientific Success

Science is successful in many ways. Scientists can treat diabetes, measure continental drift, take photographs of Mars and sequence the genome. Scientific theories give explanations of celestial events, biological diversity and economic change. For centuries scientific work has received financial support and intellectual respect. Scientific ideas have permeated literature and the arts. Science is an exemplar of a successful institution.

The above successes have different connections to the goal, or goals, of science. It is not a goal of science, although it may be a goal of scientists and of scientific institutions, to gain social support and cultural influence. Social support—at least as expressed in financial support—is a prerequisite for doing scientific work but not the goal of the work produced. Cultural influence is an occasional consequence of scientists' achievements but, again, not constitutive of the scientific achievement itself. (Similarly, healing the patient is a goal of medicine, but making money is only a goal of the health care providers.)[1]

There is more agreement about what is *not* a goal of science than about what is, or can be, a goal of science. It is a popular view that truth, significant truth, approximate truth or accurate representation is the goal of science (see, e.g., Boyd 1980, Giere 1988, Kitcher 1993, Thagard 1992). According to this view—scientific realism—successful scientific work contributes to the corpus of known truths (significant truths, approximate truths). It is almost as common, however, to doubt that scientific progress is marked by such accumulation of truth. Kuhn (1970 [1962]) began and Laudan (1981) continued historical work showing that scientific change frequently overturns previous beliefs rather than aggregates truths by building on prior work. For these scientific antirealists, truth is not an achievable goal of science. Laudan's view of scientific success is:

> A theory is successful so long as it has worked well, that is, so long as it has functioned in a variety of explanatory contexts, has led to confirmed predictions, and has been of broad explanatory scope. (Laudan 1981, p. 226)

Realists[2] agree with the content of this account of scientific success, as far as it goes. They regard it as *incomplete:* typically, realists argue that truth explains this success and is the ultimate goal of science. Truth is an *interpretation* of scientific success that realists accept and antirealists do not accept.

For the purposes of this chapter, the disagreement between realists and antirealists is less interesting than the extent of their agreement: realists and antirealists mention the same kinds of things as indicative of scientific success, namely, explanatory success and predictive success. This chapter aims to produce an account of scientific success, since Laudan's account is not definitive and there are no better ones in the literature. In particular, I will argue that empirical rather than theoretical successes are primary. The primacy of empirical success shows that, in science, the world is taken to be contingent. Chapter 3 explores the differences between realist and antirealist accounts.

An account of scientific success, and an account of the goal or goals of science whose achievement constitutes scientific success, are necessary preliminaries to providing an account of scientific rationality. Naturalistic philosophy of science approaches the question of success by asking which scientific methods, heuristics, policies and strategies are effective. Scientific rationality is thus an instrumental rationality,[3] and the goal sought is at least scientific success and perhaps also truth.

2 Empirical Success and Theoretical Success

Many kinds of results are considered constitutive of scientific success. I sort them into *empirical* and *theoretical* successes. Empirical successes include predictive, retrodictive, experimental, and some explanatory and technological successes (to be discussed below). Theoretical successes include simplicity, conservativeness, causal adequacy, consistency, elegance and breadth of scope. Typically, scientists value some selection of both empirical and theoretical successes.

The difference between empirical and theoretical success can be stated simply. There is something in common to all cases of empirical success, and unnecessary in all cases of theoretical success. That is, *empirical successes are contingent on the world outside the inquirers; theoretical successes are not.* Production of a simple theory, or an elegant theory, or a theory reflecting a particular ideology is contingent only on the ingenuity of the inquirers and the flexibility of their conceptual schemes. Theoretical success is thus due to factors internal to scientists and their theories while empirical success is due to dependable behavior by the world. It is thus straightforward to tell the difference between theoretical and empirical success on empirical grounds: one merely asks the question, does the success depend only on conceptual, psychological, etc., factors or does it also depend on the world being investigated?[4] It may be that a theoretical success is often *accompanied by* empirical success (e.g., simplicity or breadth of scope might on occasion be accompanied by predictive power, so that we have evidence that the world is simple or that it is unified); still, it is theoretical success, because a theory can have theoretical success (such as simplicity and breadth of scope) without having any empirical success.

There is variability in what counts as a theoretical success. Different historical periods, different scientific fields and different scientists all can have different—even contrary—views about theoretical success. For example, causal completeness, a theoretical standard since Greek science, was transformed in the scientific revolution into a focus on efficient causation, and ultimately rejected by most scientists who accept quantum mechanics. And, for example, European geologists (e.g., Wegener and those who took his ideas seriously) in the early part of this century valued theories with breadth of scope far more than did British and American geologists who preferred more local theorizing.

Longino (1994) associates different standards of theoretical success with the various values held by scientists. One of her best demonstrations of this is her list of "feminist theoretical virtues" which she compares with the standard list of "theoretical virtues." The standard list (taken from Kuhn 1977 and others) includes accuracy, simplicity, consistency, fruitfulness and breadth of scope. The list of "feminist theoretical virtues" includes novelty, ontological heterogeneity, mutuality of interaction, applicability to current human needs and accessibility of ideas.[5] The list comes from work on case studies, where feminist concern to reveal the influence of gender as well as produce scientific success profit from these theoretical goals.

Some of Longino's "feminist theoretical virtues"—especially novelty, ontological heterogeneity and mutuality of interaction—are clearly shown by case studies to be alternative standards of theoretical success. (The others, I think, satisfy goals other than the goals of science, such as pragmatic and political goals.) For example, Keller's (1983) study of Barbara McClintock shows the importance of ontological heterogeneity and mutuality of interaction. Profet's (1993) theory about the biological function of menstruation shows the importance of novelty. Simplicity and conservatism (Quine's 1960 list of theoretical goals of science) do not even rank in this list of the feminist theoretical virtues.

Kuhn (1977) and Longino (1990) have made much of the variability of values in science and their potential for conflict with one another. Even traditional values, such as scope and simplicity, can conflict with one another, requiring ranking of values for resolution. For both Kuhn and Longino, individual and group variability in values—*which* values and how to *weight* them against one another—invariably produces dissent among scientists. Also, they both claim that there is no point of view transcendent of these values from which to arbitrate differences; the most that can be done is to negotiate based on the values that are shared.

When looking at which values are negotiable and how they are negotiated, a definite pattern emerges. It is notable that Longino states that only "empirical adequacy" is a non-negotiable value of science; if this is not a value, the activity is not science (see 1990, p. 77).[6] "Empirical adequacy" is the only virtue of science she mentions that is *not* a theoretical virtue. Why does Longino say that "empirical adequacy" is a constitutive value of science? She is not aiming to stipulate anew goals for science; rather, she means to give an account of our best practices. Her argument for this point is not stated. My sense is that she is reflecting on her historical and contemporary knowledge of science.

It is striking that theoretical goals of science have readily been trumped by empirical success in historical cases. For example, as mentioned above, theoretical values such as causal completeness (determinacy) were given up when it was thought that the Copenhagen interpretation of quantum mechanics has singular empirical successes. And, for example, concerns about the mechanism of continental drift (the theoretical goal was explanation in terms of mechanisms) were set aside when the paleomagnetic evidence for drift became strong. As late as 1964, a well-respected geophysicist, Gordon MacDonald, argued in *Science* that it was physically impossible for there to be mantle convection currents producing seafloor spreading (see Menard 1986, p. 227). No one had much of an explanation of seafloor spreading in the 1960s, and the mechanisms are not fully known yet. None of this deterred geologists who were persuaded by mounting paleomagnetic and oceanographic evidence. They set aside MacDonald's concerns. Also, simplicity is frequently given up when phenomena prove to need a more complex account. For example, the master molecule theory of Watson and Crick is now generally seen as an oversimplified view of cellular organization, which fails to take account of the evidence for cytoplasmic inheritance, nuclear regulation and intranuclear interactions. These examples are all twentieth century cases, but the regular phenomenon of evidence disposing of theoretical values goes back at least to the Scientific Revolution when, for example, empirical success in mechanics led to discarding the Aristotelian requirement of teleological explanation.

From the perspective of naturalistic epistemology, which I take throughout this book, theoretical values are not given prior to investigation of the world. *A priori,* there is no reason to prefer simple over complex, unified over disunified, hierarchical over self-organizing representations of the world. Among naturalistic epistemologists, realists regard theoretical values as *testable presuppositions* about the world, and antirealists regard theoretical values as *pragmatic constraints* on theorizing. Realists acknowledge that the world might be simple or complex, unified or disunified (and thus explainable by theories of wide or narrow scope), hierarchical or self-organizing, and even best modeled by consistent or inconsistent theories (since humans may not be capable of exactly representing the world). Theoretical values are testable and, historically, evidence has disposed of theoretical values when they do not survive empirical tests. Thus, for the realist, theoretical successes are not only secondary to empirical success, but valued only when they bring empirical success.[7]

Antirealists claim that more than one theory can be empirically adequate. Among the empirically adequate theories, we prefer some for pragmatic reasons, such as psychological ease of use (e.g., simple and conservative theories are often psychologically easier to use). Sometimes the pragmatic considerations are broadened to encompass moral and political factors such as conduciveness to satisfying human needs and wide accessibility of ideas. Longino (1990), for example, includes such reasons for preferring one empirically adequate theory over another.

No position is taken on the realism-antirealism debate in this chapter. There is no need to settle the debate in order to talk about the difference between theoretical values and empirical success. Moreover, intermediate positions are possible with respect to the status of theoretical values. Some theoretical values could be presuppositions about the world, and others pragmatic constraints.

Variability of theoretical values is not in itself an indication that they are secondary or subordinate values of science. After all, the goals of science might not be universal. The important observation is that theoretical values are regularly relinquished for empirical success, and that there is a point of view from which to arbitrate conflicts in theoretical values, not in general (e.g., novelty will not *always* trump conservativeness), but for particular cases. Empirical success is a primary goal of scientific inquiry, and theoretical success valuable only when it brings extra empirical success, convenience or moral benefits with the available empirical success.

Empirical success may not be the only primary goal of science. Thus, I describe it as "a primary goal." Truth, representation or some variant of these may turn out to be a primary goal of science also.

Theoretical values are indispensable for science, just as theorizing in general is indispensable. I am not arguing that scientists should discard theoretical values and aim only for empirical success. The requirement to aim for more and more empirical successes is not specific enough to direct research. Theoretical values function as heuristics drawn from plausible theory or pragmatic goals. Scientists need all the heuristics they can get.

3 Defining Empirical Success

No one has a good, comprehensive definition of empirical success. Several writers have insightfully described one or another aspect of empirical success. Hacking (1983) celebrates manipulative success, Haraway (1991) notes that nature often surprises us and Latour, Pickering and others write of the "resistance of nature" to which we must accommodate when producing empirical success. More traditional philosophers of science, such as Van Fraassen (1980) continue to talk of "saving the phenomena," understanding empirical success in terms of observational and predictive success.

Following the constructivist excesses of sociology of scientific knowledge, there is a new acknowledgment in science studies that empirical success is not manufactured wholly by humans and their instruments, but depends on the world. The main task of this chapter is to give such an account of empirical success that is also adequate for the epistemology of science that will follow. Three central issues will be: first, describing the *variety* of empirical successes; second, exploring what counts as a *robust* empirical success and third, discussing the notion of *significant* empirical success.

The variety of empirical successes

Empirical successes can be observational, predictive, retrodictive, experimental, explanatory or technological—to mention some of the main categories. During different historical periods, and in different fields, there has been focus on one or another kind of empirical success. For example, during the Scientific Revolution of the seventeenth century, an ardent interest in experiment developed. In the field sciences of the nineteenth century (geology, evolutionary biology) the emphasis was on giving a unified explanation of observations. Much of empirical success in paleontology is explanatory. Some of recent geology (e.g., paleomagnetism) is predictive of observations. Astrophysics both predicts and retrodicts observations. Molecular biology and particle physics are highly experimental. Understanding of the action of insulin and other drugs (e.g., Tamoxifen, Pergonal, human growth hormone) depended on technological success in purifying extracts of living tissues.

Sometimes the focus on a category of empirical success is due to preference and sometimes to necessity. For example, during the Scientific Revolution there was preference for experimental success because of the general growth of technology at that time. Intervention in nature, or "twisting the lion's tail" as Francis Bacon called it, was generally valued because it was needed for the advances in navigation, transportation and military technology at that time. Other kinds of empirical success (such as more passive observational success) were available, but not valued as much by the leaders of the Revolution. On the other hand, for example, empirical success in paleontology is limited, of necessity, to explanation of observations and vague predictions. Not enough is known about (or even, could be known about) genetic mutation and environmental variations to make anything other than the vaguest of predictions about evolutionary change. And, obviously, experiments on evolutionary change are not possible for most organisms.[8]

There is no "paradigmatic" or "typical" kind of empirical success. This is another instance of the disunity of science. Attempts to define "empirical success" (such as the ones mentioned above by Hacking, Van Fraassen, etc.) run the risk of being too specific, and not including all empirical successes, or too vague, and not characterizing anything. I will ground my discussion in this section by describing a range of examples of robust and significant empirical success. In later sections, a contrasting range of examples where there is no empirical success, or limited empirical success, will be discussed.

Discovery of insulin Over a year from 1921–1922, insulin was isolated and its physiological properties demonstrated by a group of researchers at the University of Toronto: Frederick Banting, Charles Best, J. B. Collip and J. J. R. McLeod. The result was a cure for diabetes. The discovery of insulin came through a lengthy series of experiments, first on dogs and then on other animals and finally humans. There were failures and equivocal results as well as successes, and much trial-and-error in experimental design. Michael Bliss has documented the progress of the research, as he says, "dog by dog, day by day, experiment by experiment" (1982, p. 17).

Pancreatic extracts caused toxic reactions, so experimenters tried various ways of isolating the endocrine secretions, beginning with ligating pancreatic ducts in order to atrophy the exocrine function, and then removing the pancreas weeks later and preparing an extract for use on a depancreatized (i.e., diabetic) dog. Sometimes such an extract appeared to work, other times it appeared not to work, and sometimes there was a toxic reaction. At the same time, the researchers were improving their measurements of blood sugar. The experiments became easier when they started using fetal pancreases, which did not need prior ligation. They kept a depancreatized dog alive for about 6 weeks with injections of pancreatic extract. Paradoxically, for they did not yet know about insulin shock, the dog died after a convulsive reaction to insulin injections.

In December 1921, Collip was assigned the task of purifying and standardizing the extract, so that its physiological properties could be clearly demonstrated. Collip described his methods as "bathtub chemistry" (Bliss 1982, p. 116). He tinkered with solvents, trying to isolate the active principle. In late January 1922, he succeeded with alcohol as a solvent, and tested its potency on rabbits. Collip showed that the extract abolished ketosis and enabled the liver to form glycogen. In February insulin was used successfully on human diabetics. Then, unexpectedly, Collip lost the knack of producing insulin. There was an "insulin famine" until mid-May, when Collip succeeded again, this time using acidified acetone as a solvent. Soon afterwards, the production of insulin was taken over by Eli Lilly and Company. Bliss (1982, p. 127) comments: "It is probably impossible to specify one time . . . when it could be said that insulin had been discovered."

This is an example of an experimental and technological success. The success began early in the fall of 1921. At first it was impossible to be sure that the results were robust. Success increased in fits and starts, until by the summer of 1922 it was definite and striking. It is notable that the success depended on tinkering, rather than on explicit and reasoned redesign of procedures.

Cases of similar kinds of empirical success are the development of the anthrax vaccine (see Latour 1988) and the beginnings of Mendelian genetics (see Kohler 1994).

Magnetic symmetry patterns and seafloor spreading The Vine-Matthews hypothesis, first put forward in a 1963 paper, suggests that the magnetic stripes found parallel to oceanic ridges are produced by slow seafloor spreading. With a spreading rate of a few centimeters a year, stripes are expected as a result of reversals in the earth's magnetic field. In 1963, explanations and predictions of magnetic striping were qualitative in nature and not related to other measurements of paleomagnetism. Over the next three years, Vine, Matthews, Tuzo Wilson and several others used the Vine-Matthews hypothesis to give quantitative predictions and explanations of observations that were so successful and wide ranging that some even thought for a time that the data was "too perfect" (e.g., Joe Worzel, quoted in Glenn 1982, p. 335).

Vine and Matthews's first explanations were of the Carlsberg Ridge in the Indian Ocean. They showed that three traverses of the ridge gave similar magnetic patterns and explained this in terms of seafloor spreading. In 1965, Wilson realized that the Vine-Matthews hypothesis predicted a symmetrical pattern on either side of the ridge crest, and then found this pattern across the Juan de Fuca ridge in the Pacific Ocean. Vine and Wilson also showed that the magnetic reversal chronology developed from land-based measurements (volcano lava) corresponded to the magnetic patterns across the Juan de Fuca ridge, if a constant spreading rate was assumed (a reasonable assumption). They found the same symmetry pattern and correspondence with the magnetic reversal time scale for the East Pacific Rise. As more details of the magnetic reversal chronology were discovered (e.g., the Jaramillo event), the correspondences grew even stronger.

Lamont laboratory (at Columbia University) surveys, analyzed in 1965, over the Reykjanes Ridge near Iceland and over the East Pacific Rise showed not only symmetry and correspondence with the magnetic reversal chronology but also a match of the ridge patterns with one another, once different spreading rates were taken into account (1 cm/year for the Reykjanes Ridge and 4.5 cm/year for the East Pacific Rise). At this point, if not earlier, the empirical success of the Vine-Matthews hypothesis was inescapable. Moreover the predictive and explanatory successes continued, being extended first to the Carlsberg and the Red Sea rift, and then worldwide. Soon afterwards, data from the Deep Sea Drilling Project, which analyzed remnant magnetism in sediment cores, showed (by 1970) the predicted correspondence between magnetic patterns in the cores (perpendicular to the Earth's surface), magnetic patterns across ridges (parallel to the Earth's surface) and the magnetic reversal chronology developed from terrestrial measurements. Tuzo Wilson remarked:

> Three different features of the Earth all change in exactly the same ratios. These ratios are the same in all parts of the world. The results from one set are thus being used to make precise numerical predictions about all the sets in all parts of the world. (Quoted in Glen 1982, p. 351)

To be sure, not all the data were perfect. Some traverses of ridges produced magnetic patterns that were less than symmetrical. Before the Jaramillo event was discovered, there was not an exact match between seafloor and terrestrial measurement. However, considering the number of uncertainties and variables affecting measurements of remnant magnetism (e.g., differences in the thickness and composition of ocean floor, measurement errors, partial sampling of volcanic rocks), this was not surprising. What *was* surprising, and impossible to dismiss as due to chance, were the symmetries and correspondences in the best data.

Note that these empirical successes are predictions of observations, and explanations of observations. No experiments or manipulations of phenomena were performed. Empirical successes in astrophysics are similar in kind.

Evolutionary biology in the late nineteenth century[9] Darwin's *Origin of Species* (1859) persuaded many that there is transmutation of species. The biogeographical data that Darwin supplied, many from the *Beagle* voyage, displayed local adaptations that are not (except in an ad hoc manner) explainable in any other way. Darwin's particular explanation of the mechanism of species change—natural selection—was not, however, widely embraced. Several theories, some in combination with one another, were used, successfully, to explain the variety of biogeographical, paleontological, embryological and other data. Some successful, albeit rather vague, predictions were also made.

Both natural selection and Lamarckism were used to explain adaptations of species to their environments. Both explain and predict continuous variation. Both successfully predict that there will be more findings of adaptation (although they do not specify exactly *how* species will be adapted or to what *degree* they will be adapted). Both predict that intermediate forms between species will be found in the fossil record; there was only limited success in finding intermediate forms by the end of the nineteenth century. Lamarckism in particular failed the predictive tests begun by Galton and continued by others. Darwin, of course, claimed that natural selection on random variation is the mechanism of species change, although, in later publications, he also thought that the environment produces heritable species change. His theory of pangenesis was given as a mechanistic explanation of Lamarckian species change. A few biologists (such as Wallace, Hooker and Weismann) were even more impressed with natural selection as a mechanism of species change than was Darwin. But most thought that natural selection is not a powerful enough process to account for species change. Among them were Lamarckists, such as Haeckel.

Many thought that gradual and adaptive changes, whether produced by natural selection or Lamarckian mechanisms, cannot gather enough momentum to avoid being "washed out" by random breeding in successive generations. The mutation theory (saltationism) and orthogenesis were two theories that denied that evolutionary change is gradual and adaptive.

According to the mutation theory, espoused by Galton and Huxley, species can show huge variations in just one generation. Sometimes (e.g., with Huxley), the mutation theory went together with natural selection as an account of species change. The mutation theory explains observed phenomena such as the occasional occurrence of hexadactyl humans and Ancon sheep. It explains the absence of intermediate fossil forms. It predicts, nonspecifically, other saltative mutations, and these predictions were not confirmed in the nineteenth century.[10]

Orthogenesis—the evolution of species under the direction of internal forces—was very popular. It explains some fossil findings of species with apparently non-adaptive modifications of earlier species, such as intermediate stages of the evolution of the horse, and later stages of the evolution of cephalopods into degenerately large forms. Mivart, Owen, Cope and Hyatt were prominent developers of orthogenesis. Orthogenesis predicts that evolutionary change will generally be non-adaptive, and also predicts parallel evolution (e.g., that the reptiles of today will evolve into the birds and mammals of the future, just as happened in the evolutionary past).

Finally, correspondences between embryological stages and evolutionary stage were explained by Haeckel's thesis "embryology recapitulates phylogeny." Both Lamarckists and orthogenesists saw in the recapitulation theory possible mechanisms for species change, and both could explain the observations of recapitulation. They hypothesized that embryological development evolved by adding additional stages, either repeating the adaptive growth of adult ancestors, or by internal development.

Most of the empirical successes in nineteenth century evolutionary biology were explanations of observations (paleontological evidence). There were a few successful predictions, although the predictions were much more vague than, e.g., the predictions of findings in paleomagnetism in the 1960s.

So far, I have not defined empirical success. The above examples show that empirical success is not completely captured by any one of the well-known paradigms of empirical success: predictive accuracy (Van Fraassen's paradigm), explanatory power (Laudan), technological success (Latour) or manipulative success (Hacking). To be sure, particular empirical successes involve one or more of these. Empirical success can be achieved explicitly or non-explicitly; the latter depends on how much non-linguistic manipulation (e.g., tinkering with equipment and procedures) is involved.

It is my view that general definitions of empirical success say little, and that understanding and characterization of empirical success comes from looking at types of examples, as I have just done. But general definitions can be given, for example: when there is empirical success, scientists, instruments and the world successfully coordinate their actions as a result of tinkering, conceptual adjustments and serendipity.[11]

Robustness

In the three examples above—discovery of insulin, confirmation of seafloor spreading and evidence for evolutionary theories in the nineteenth century—most empirical success was shown to be robust over time, as more experiments, observations and explanations were generated. Often, however, empirical success is not dependably produced. Experiments fail to replicate, or data is inaccurately analyzed, or observations are partial and misleading. For example, Pons and Fleischmann's claim to have obtained excess heat in electrolytic cells with palladium electrodes was not dependably reproduced (see chapter 7). Yasargil's claim in the late 1970s to have reduced strokes by use of a surgical technique, the EC-IC bypass, was based on poorly analyzed data (see chapter 7). Marsh and Cope claimed to find a linear process of evolution of the modern horse in the paleontological evidence, but subsequent fossil evidence did not fit into this linear pattern.

While it takes time to determine whether an empirical success is robust, typically it takes less time, and fewer considerations, than does assessing a whole theory. Moreover, determination of empirical successfulness is *separable* from theoretical disputes. For example, Judah Folkman claims that angiostatins (substances that inhibit the growth of blood vessels) can generally control solid tumors. In 1998, he reported that he successfully used genetically engineered angiostatin to arrest the growth of tumors in laboratory mice. At first, other researchers were unable to replicate his results and the empirical success was in question. About a year later, it was discovered in Folkman's laboratory that zinc is essential to this genetic engineering, and the experiment was reliably replicated in another laboratory.[12] The jury is still out on the theoretical dispute about whether angiostatins will be generally useful in cancer therapy.

If robustness is not a requirement—for example, if data are theory laden by the theories under consideration, or if data are sloppily gathered—then empirical success would not be surprising or creditworthy. In the examples given, scientists counted empirical success only when it was carefully gathered and not theory laden by the theories under investigation, and they went through reasonable precautions to ensure this. Those in the SSK tradition who claim that *all* data are theory laden by the theories under consideration are taking an extreme skeptical position that is devoid of content, since they do not say what would count as data *not* laden by the theory under consideration.

Even Kuhn, who did more than anyone to draw attention to "theory ladenness of observation" and "incommensurability," stated more recently that there are "concrete technical results" for which "little or no translation is required" and which can be critical for producing scientific change (1977, p. 339). Kuhn went on to specify that these "concrete technical results" consist of predictive accuracy and fruitfulness (of new phenomena or new relationships between phenomena (1977, p. 322)), and perhaps also scope, but not consistency and simplicity (1977, p. 339). Interestingly, this classification corresponds to my distinction between empirical and theoretical success.[13] And the word "concrete," which Kuhn himself does not specify further, is clearly interchangeable with "robust" in my usage.

Significance

A theory that correctly predicts only that the sun will rise tomorrow does not have significant empirical success. A theory (such as Freud's theory of the psyche) that can explain almost any behavior, but predicts little, does not have significant empirical success. Monday morning quarterbacking is unimpressive. Empirical successes are significant when they are mostly attributable to the theory, rather than to prior knowledge shaping the application of a theory. Such significance is clearly produced when a theory has *new* empirical success, i.e., new phenomena are discovered or produced, or phenomena are explained for the first time. The more precision that this is done with, the greater the empirical success.

Significance is a *qualitative* measure of empirical success. Fortunately—because producing a quantitative measure would be onerous and perhaps impossible—this is all that is needed for this epistemology of science (as I will show in later chapters).

Explanation is unique in its range of degrees of empirical success. At one extreme, some kinds of explanation do not even count as empirical successes at all. For example, the electromagnetic ether explains how light can be a wave (to those who think that waves need a medium), but it does not explain *phenomena.* The electromagnetic ether provides an explanation of a theoretical claim. It is an example of a theoretical success rather than an empirical success. At another extreme, some kinds of explanation are highly empirically successful. For example, Vine, Matthews and Wilson explained the magnetic anomalies parallel to oceanic ridges in terms of seafloor spreading. They even explained, with quantitative accuracy, details of the magnetic patterns that were not even noticed before, such as the bilateral symmetry and the correspondences with other oceanic ridges and with magnetic reversal chronologies.

Many explanatory successes fall between these extremes. For example, the cases of explanation in nineteenth century evolutionary biology mentioned above are empirical successes, because they explain paleontological data, with some significance since they could not explain just anything. Adaptive theories of evolution (natural selection and Lamarckism) explain a wide range of data, including data discovered after the theories are proposed. They are even weakly predictive. Orthogenesis led to noticing, as well as explaining, linear patterns and non-adaptive structures in the fossil record (much of this evidence failed to be robust). Less significance is typically found in sociobiological explanations of human behavior, often judged to be "just so" explanations (see, e.g., Kitcher 1985). As social psychologists have discovered (see, e.g., Nisbett and Ross 1980, chapter 6), it is fairly easy to produce a causal explanation of a phenomenon in hindsight, and such explanations are often psychologically satisfying. Yet they accomplish little because it is not the theory (the causal explanation) that is responsible for the explanatory success; rather, it is prior knowledge of the phenomenon. This does not yield more explanatory success, and, in particular, it does not yield explanatory success of new (previously not observed or not noticed) phenomena.

Technological successes are always empirical successes, but some are more significant, in the sense meant in this section, than others. Some technological successes produce novel phenomena, while others are unsurprising applications of noncontroversial theories. For example, design of the side-by-side refrigerator-freezer did not require novel theorization or yield new kinds of phenomena, although it was a creative technological success. The discovery of insulin, on the other hand, as described above, required novel theorization and produced novel phenomena. No precise demarcation is needed here; but this is where I find a useful distinction between science (including some technoscience) and *mere* technology.

Predictive success, whether predictive of experiments or observations, likewise has a range of significance, also depending on the novelty and quantity of prediction. Predictions can also be more or less precise, ranging from the vague prediction, for example, that species will be adapted to their environment (which does not say how they will be adapted, or to what degree) to the more precise prediction that, for example, there will be a lunar eclipse of the sun in Cornwall, United Kingdom, on August 11 1999 at 11 am.

Empirical success is an essential product of scientific work. Theoretical goals of scientific work are secondary. This chapter was devoted to describing the nature of empirical success with a level of detail that will be important in succeeding chapters. There are different types of empirical success—observational, predictive, retrodictive, experimental, some explanatory and some technological—supporting the thesis of the disunity of science. Two measures of empirical success that will be useful, beginning in the next chapter, are *robustness* and *significance.*

3
Whig Realism

Too much Truth
Is uncouth

Franklin Adams

1 Introduction

The goal of this chapter is to present a new version of scientific realism, which I call *whig realism.* Whig realism gives a novel explanation of empirical success in science. This chapter also takes the discussion of scientific goals further than in chapter 2. Truth, in addition to empirical success, is a primary goal of science. As a consequence, whig realism has methodological import.

The term "whig" is often used pejoratively in science studies. This started with Herbert Butterfield's essay, *The Whig Interpretation of History* (1965 [1931]). Butterfield urged historians of science to avoid "the historian's 'pathetic fallacy' . . . [of] organizing the historical story by a system of direct reference to the present" (1965, pp. 31–32). Butterfield took the term "whig" from the British political party of nineteenth century Protestant gentlemen who saw the present as the culmination of progressive historical trends. I will argue that, while whig history is indefensible as history,[1] whig realism is a plausible philosophical position.

Whig realism shares with whig history the practice of assessing truth in past theories from the perspective of present knowledge. It differs in that it does not *generally* impose the perspective of the present on the past. Whig realism will be developed and defined precisely, after I make some observations about the current literature on scientific realism.

2 Scientific Realists and Antirealists

The philosophical debate over scientific realism has been lively for about a century, since the work of Mach, Duhem and Poincare.[2] Attention is always given to the ways in which realists and antirealists differ in their interpretations of science. The ways in which they agree, however, are telling.

As mentioned in chapter 2, realists and antirealists *alike* are impressed by the empirical successes of science. They *disagree* about whether truth (or partial truth, approximate truth, correspondence, representation, etc.) explains robust, significant empirical success. Putnam famously wrote "The positive argument for realism is that it is the only philosophy that does not make the success of science a miracle" (1975, p. 73); later writers (e.g., Boyd, Kitcher, Giere) have made use of this or similar arguments. Antirealists, on the other hand, have argued that truth (or partial truth, approximate truth, correspondence, representation, etc.) cannot explain success because such an explanation would be circular (Fine 1986), not needed (Fine 1986 and Van Fraassen 1980) or not explanatory because the explanans is false (Laudan 1981).

Not only is there general agreement that scientific work is often empirically successful, there is also frequently agreement about *which* scientific achievements are empirically successful. This is not surprising, given what was said in the previous chapter about robustness of some empirical successes. For example, there is widespread agreement that Ptolemaic astronomy, the theory of circular inertia, Galenic medicine, Newtonian mechanics, the phlogiston theory, Maxwellian electromagnetism, Lamarckian evolution, Skinnerian behaviorism and classical (linear, formal) AI in cognitive science[3] are each empirically successful. These examples are deliberately taken from Laudan's list of empirically successful yet (according to Laudan) false theories (Laudan 1981, pp. 232–3), and supplemented with similar examples. So realists and antirealists agree that science is often empirically successful, and even agree about which theories are empirically successful. Moreover, judgments of empirical success, for both realists and antirealists, are *independent* of judgments of truth. Realists disagree with antirealists about the veracity of past theories (how much truth) and about the proportion of past theories that are false but not about which theories are successful. For the realist, truth (or something similar) explains the majority of successes; for the antirealist, truth does not explain success. Either way, it is frequently judged that there is empirical success.[4]

Given the agreement of judgments about empirical success, it is surprising that there is no generally accepted definition of empirical success. That is why I did some work in that direction in chapter 2. Chapter 2 establishes a distinction between empirical success and other kinds of success—for example, theoretical virtues (such as elegance or simplicity) and rhetorical factors (i.e., those aspects of a theory or practice that make it psychologically, ideologically or politically appealing). Chapter 2 also discussed the characteristics of different cases of empirical success: examples of observational success, predictive success, experimental success and some kinds of explanatory and technological success. Empirical successes have in common that they are all contingent on the part of the world under investigation. They are evaluated in terms of their robustness and significance.

Does truth (approximate truth, partial truth, correspondence, etc.) explain empirical success?[5] Only if empirical success is robust, significant enough (sufficiently novel and abundant to be in need of explanation), and truth is a good explanation. Antirealists argue against both of these claims. Van Fraassen (1980, p. 40) and Fine (1986, chapter 7) argue that there is nothing about the empirical successes of science that is in need of explanation: theories are selected for their past success and generally show no extra success after that selection. They deny that theories have novel empirical successes at anything more than an accidental rate. Laudan (1981) uses historical examples, such as the ones mentioned above, to show that successful theories are false often enough that truth cannot explain success.

Van Fraassen's and Fine's argument has received little attention. It is inappropriate to conduct a *general* discussion of the argument, since the argument depends on cases and thus depends on data about the results of chosen theories. Van Fraassen and Fine typically assert that theories are successful at only chance rates outside the applications in which their success first impressed scientists. Also, they claim that conjoining theories results in no more successes than one would expect on grounds of logic and chance. Data in support of these assertions is in short supply at present, and Van Fraassen and Fine do not contribute to it. It is easy to cite contrary data. The case of empirical success of the seafloor spreading hypothesis is a counterexample, for it is a clear case of continued significant empirical successes of a theory that cannot reasonably be dismissed as due to fortuitous extension of theory. The case of the empirically successful conjoining of Mendelism and Darwinism in the 1940s is another counterexample. More counterexamples, of course, are needed to fully counter Van Fraassen's and Fine's arguments.

Laudan's argument against scientific realism has received the most attention. Recently, Kitcher (1993) has developed a sustained argument for realism, using the same set of historical examples that Laudan focuses on and countering Laudan's argument. Kitcher's ideas are the springboard for my own views, and I will discuss them first, coming back to Van Fraassen's and Fine's arguments at the end of the chapter.

3 Kitcher's Recent Defense of Realism[6]

Kitcher's *The Advancement of Science* (1993) acknowledges that scientific theories often contain falsehoods that are later abandoned. Yet Kitcher maintains that, on balance, scientific progress is veridical progress (i.e., increase in overall truth content of theories). Moreover, he argues that scientific progress is marked by aggregation of truths to a core, firm set of truths. The crux of the argument is Kitcher's attempt to measure the veridicality of a theory in a new and more sophisticated way. He has a new theory of reference and a new view of the essential core of a theory. According to these new accounts, terms like "phlogiston" sometimes refer, and the essential core of successful theories—comprised of the "working posits"—is true. Let me explain these ideas as Kitcher explains them, with his examples.

Kitcher says of Priestley's work: "Inside his misbegotten and inadequate language are some important truths about chemical reactions, trying to get out" (1993, p. 99). Laudan rejects such statements because he says (with many others) that "phlogiston" fails to refer and that therefore phlogiston theory is essentially false because existential claims with a non-referential subject are false. Kitcher criticizes the implicit theory of reference that Laudan and others use (i.e., that reference is fixed by a description, in this case "phlogiston" is "the substance emitted during combustion") and suggests in its place a theory of "heterogeneous reference potential" according to which terms have the potential to refer in several different ways when scientists already use them in several different contexts. "Phlogiston" will turn out to have reference at least sometimes because "dephlogisticated air" was successfully—*albeit unwittingly*—used in a particular set of experimental contexts to refer to oxygen (Kitcher 1993, p. 100). That is, "phlogiston" sometimes refers even though there is no such thing as Priestley's phlogiston, and even though no phlogiston theorist at the time could say how the term refers. Phlogiston theorists were actually often successfully referring to oxygen when they used the phrase "dephlogisticated air," although they did not know it.

Kitcher also argues that we should distinguish between "working posits" and "presuppositional posits" of a theory: stability of the "working posits," he claims, is the only concern in discussions of realism. The electromagnetic ether, as developed by Fresnel and Maxwell was, according to Kitcher, a mere presuppositional posit, "rarely employed in explanation or prediction, never subjected to empirical measurement" (1993, p. 149). For Kitcher, as for many, explanatory, predictive and observational success constitute scientific success. The posit of an electromagnetic ether is not involved in the production of scientific success of electromagnetic theory, so it may be discarded without changing the essential working core of the theory. The explanation of success of electromagnetic theory can then lie in the truth of that core of "working posits."

Kitcher's conclusion is that Laudan's arguments leave "untouched the central realist claim. . . . Where we are successful our references and our claims tend to survive. . . . Our successful schemata employ terms that genuinely refer, claims that are (at least approximately) true" (1993, p. 149). Scientific realism is reinstated.

4 Discussion of Kitcher's Views

Let's make use of Kitcher's notions of reference potentials and presuppositional vs. working posits and explore his historical examples further.

Priestley had no idea when or how "phlogiston" in fact refers, nor did Fresnel and Maxwell have any idea that ether was a presuppositional rather than a working posit. For Priestley, the idea that burning objects give off a substance which he called phlogiston was the cornerstone of his theory of combustion. For Fresnel and Maxwell, ether was indispensable for the development, statement and plausibility of electromagnetic theory. Ether was modeled by an analogy with fluid dynamics (see, e.g., Nersessian 1992). In hindsight, ether may have been logically dispensable for prediction and unnecessary for explanation of observations, but, for Maxwell and other physicists at that time, ether was *psychologically, historically* and *explanatorily* indispensable.

Most realists—Kitcher included—take realism to imply that our successful theories are substantially right. Indeed, Kitcher thinks that the core of truths grows as theories develop and change, and that this is the mark of scientific progress. However, according to Kitcher's views about reference and presuppositional posits, it cannot be concluded that our currently successful theories are substantially true or referential *in anything like the ways that we state and understand them.*

Only in hindsight can it be said how "phlogiston" referred, or how the posit of the electromagnetic ether can be omitted from mature electromagnetic theory. Thus, only in hindsight might we be able to say whether there are, for example, quarks, and which of our theoretical claims are central and true. This is not traditional realism. In fact, it sounds closer to Laudan's antirealism. After all, Laudan's historical claim is that the parts of a successful theory that we take to be central and true can be abandoned in later work.

5 Whig Realism

As a matter of fact, I don't think that Laudan-type antirealism is the only conclusion to draw from these reflections. There is a kind of realism that can be defended, which I call *whig realism.* Whig realism is a development of Kitcher's ideas, especially his ideas about reference and about presuppositional vs. working posits. Kitcher does not take his arguments to this conclusion, perhaps because there are some consequences he is unwilling to accept.

Whig realism is the only version of realism I can think of that has a chance of being true. Roughly stated, it is the position that when empirical success needs explanation (that is, when it cannot be attributed to chance or intentional choice[7]), it is due to there being *some truth in the theories.* Kitcher's phrase quoted above, "Inside his [Priestley's] misbegotten and inadequate language are some important new truths about chemical reactions, trying to get out" (1993, p. 99) is, I think, almost a statement of whig realism (it is also a whiggish statement in itself). It is not that statements made in phlogiston theory are true—indeed, most or all of them are false. But there is truth *in* the theory.

How can there be "truth in the theory" while the statements of a theory are false? The phrase "truth in the theory" is vague and metaphorical (at least I do not go as far as Kitcher, who suggests that the truths are "trying to get out"!). I'll make the idea as precise as I can:

> What is true about our empirically successful theories is, typically, an implication of the theory at the theoretical level (i.e., not just a prediction or observation) that may or may not be explicitly derived during the historical period in which the theory is accepted. Such truth can explain the empirical success of the theories.

So, for example, what is true about phlogiston theory is that combustion and respiration are dependent on the same invisible characteristics of the ambient atmosphere, and that combustion and respiration alter the composition of the local atmosphere. Priestley would have agreed to this—it is a consequence of phlogiston theory—although he would probably have regarded it as vague, and may not have made this vague statement himself. Phlogiston theory was empirically successful. In particular, it showed that some atmospheres especially support combustion and respiration (i.e., those produced through heating an oxide of mercury), some atmospheres extinguish combustion (i.e., those produced through combustion itself), and some atmospheres when combined form water (i.e., those produced through heating an oxide of mercury and "inflammable air"). These empirical successes cannot be explained by phlogiston theory. Phlogiston theory is false—and false theories cannot explain success. However, phlogiston theory has theoretical structures (they may be substructures of the official theory) that explain its empirical successes and are true. For example, it is because combustion needs particular atmospheric support that some atmospheres were found to support it, and some to extinguish it. At the time that phlogiston theory was proposed, it was not known which theoretical structures of phlogiston theory are true, but it was known afterwards, once the oxygen theory was established. This is whig realism.

The case of electromagnetic theory is understood similarly. As Kitcher himself showed, the empirical success of electromagnetic theory is due to its claim that there are fields of force acting on electrically and magnetically charged bodies in particular rule-governed ways. The posit of the electromagnetic ether does *not* explain why the theory was empirically successful—even though the theory itself would not have been developed without it.

To give a third example, the empirical success of Newtonian mechanics is explained by the fact that its laws are approximately true at low velocities. Note that this is not a truth in Newtonian theory that was derived at the time, in the seventeenth century. It is only from a whig perspective—special relativity—that motion at "low velocities" becomes a meaningful subset of mechanical phenomena, and we explain the success of Newtonian mechanics.[8]

Some more clarification of whig realism is in order here:

- Whig realism is not the view that successful theories are "approximately true." To say that a theory is "approximately true" suggests that the important, central claims of a theory are true or almost true, and some of the details are wrong. This was not the case with many theories (e.g., the examples above) even though there was "something true in them."
- Whig realism is not the view that successful theories are "partly true." To say that a theory is "partly true" is to suggest that some statements, or models, of the theory are true. Sometimes this is the case, but, often enough, what is true *in* a theory is not a subset of the statements, or models of the mature theory (e.g., phlogiston theory), and sometimes it was not even formulated at the time (e.g., Newtonian mechanics).
- Whig realism is not whig relativism. Whig relativism would be the view that we may choose to explain the success of past theories from the perspective of current theories, but those explanations will change radically

if our theories change again. So, the whig relativist denies that empirical success is really explained by (absolute) truth in a theory, because we can never say with confidence where the (absolute) truth in a theory actually lies. Whig realism and whig relativism can be distinguished in practice. The whig realist expects that, even if theories change, the explanation of past empirical success will be *similar,* pinpointing the *same* theoretical substructures as responsible for the empirical success. Moreover, the whig realist only uses empirically successful current theories—in fact, those that are more empirically successful than the past theories under consideration—for explanation of past empirical success.

• Whig realism is not skepticism. Laudan's "induction on the history of science" has led many to be skeptical of *all* truth claims in science. I am cautious about how far the "induction" extends. In keeping with the commonsense, or "homely," attitude towards truth in science (mentioned in the next part), I judge that there are many truths in science—from Harvey's claim that the blood circulates the body and the lungs powered by a four chambered heart, to the claim that there are electrons (pace Hacking) to the claim that the continents have drifted.

• Whig realism is not Fine's Natural Ontological Attitude (NOA) (see Fine 1986), although it shares with NOA a "homely" attitude towards truth in science, eschewing metaphysical theories of truth. One way in which whig realism differs from NOA is in not regarding scientists' judgments of truth as authoritative. Other differences will be mentioned later.

• Fine (1986, chapter 7) has argued that there is no general argument from the success of theories to their truth. Whig realism does not employ that kind of argument. Instead, particular truths in the theories explain particular successes. It is not the case that all empirical success is explained by truth: some successes are just due to luck, or are a consequence of past theory choice (see footnote 7). Whig realism depends on particular kinds of success—the empirical successes of new applications of a theory and of conjunctions of different theories. We do not (yet) have a formal method for deciding when a success is significant. We use plausible informal heuristics, such as: the weaker an empirical success, the more likely it is to be due to luck and not in need of explanation, and the stronger a success (in terms of range, precision, etc.) the less likely it is to be simply due to luck. Explanatory successes are often weak, since they are often *ex post facto* and frequently obviously underdetermined. Predictive successes tend to be strong, especially if what is predicted is new, unanticipated or even counter indicated before using the theory to make the prediction.

• Whig realism is consistent with a *pluralist* realism. There can be truth in more than one theory about a domain. The theories can even be inconsistent with one another without threatening whig realism, when some of the claims in the theories are not true. One important consequence of this is that consensus is unnecessary for scientific progress towards truth. Examples of this will be given below.

• Whig realism is not whig history. The whig science historian evaluates past theories in terms of their resemblance to present knowledge. The whig realist evaluates past theories in terms of their empirical success at the time they were developed; only the *explanation* of that success is ascertained from a whig perspective.

6 Evidence for Whig Realism

Whig realism is supported by historical examples in which empirical success is due to "truth in the successful theory," in the sense defined above. If there are enough such examples, whig realism is a useful philosophical position. Some examples were given above; I give more examples in this section. I take the examples from several sciences: evolutionary biology at the end of the nineteenth century, genetics before the discovery of DNA and the debate over Wegener's theory of continental drift. The examples also show the consistency of whig realism with a pluralist, rather than a consensus, account of successful scientific change.

As discussed in the previous section, much turns on the difference between a significant empirical success and one that is simply due to chance or to past choices. At present, we do not have a quantitative method of making judgments about how significant a success is. We make informal judgments. While these judgments may be imprecise, they are not vacuous. There is all the difference in the world, for example, between the successes of Vine and Matthews in predicting magnetic symmetry patterns and the explanatory successes of human sociobiology. For the purposes of this book, we do not need a precise, quantitative measure of significance. The following examples show that whig realism is a reasonable explanation only of *significant* empirical success.

Evolutionary biology during the "eclipse of Darwinism" period
This case was described in chapter 2 and will only be summarized here. During the late nineteenth and early twentieth centuries, there was a period that Julian Huxley later referred to as the "eclipse of Darwinism."[9] While it was generally accepted that there is transmutation of species ("evolutionism"), and *The Origin of Species* (1859) supplied a vast collection of persuasive evidence for this, there was lack of consensus about how the transmutation occurs. Darwin's theory of gradual evolution by means of small random variations and natural selection was only one of a number of theories about transmutation of species. Other, more popular theories were saltationism (discontinuous variation), Lamarckism (the inheritance of acquired characteristics) and orthogenesis (evolution of species under the direction of internal forces). Each theory had some empirical successes. There were examples of saltative species change, such as Ancon sheep and hexadactyl humans. There were numerous cases of adaptive evolutionary change, explained by both Lamarckism and natural selection. And, in support of orthogenesis, there was fossil evidence of parallel evolution and non-adaptive evolutionary change, such as the evolution of mammals from reptiles, the evolution of the horse and the degeneration of cephalopods. Correspondences between embryological stages and evolutionary stages were explained by Haeckel's thesis "embryology recapitulates phylogeny," and lent support to both Lamarckism and orthogenesis.

These empirical successes were typically explanations, retrodictions and predictions about observations of fossils or live organisms. In many cases, but not all, truth in the various theories was responsible for their empirical success:

Both natural selection and Lamarckian accounts were right in thinking that evolutionary change is often achieved by adaptation to the environment. This theoretical claim (evolution by adaptation) successfully explained and weakly predicted the same fossil and ecological data. The different mechanisms that they proposed for evolutionary change were not, at that time, responsible for significant empirical successes.[10] There were truths in the other evolutionary theories, too, which explain some of their empirical successes. Galton (and other saltationists, e.g., Huxley) was right in thinking that variations are not always small and therefore species change is not invariably gradual. This explains the success in noticing new, large variations. The explanatory success of Haeckel's view that "embryology recapitulates phylogeny" is also explained by the truth in it. The view is, as a general claim, false, since many stages of embryonic development do not coincide with phylogenetic histories, and vice versa. There is, however, a striking correspondence between some stages of embryonic development and some portions of phylogenetic history, and it is implausible to attribute this to chance. The more recent observations (by, e.g., Gould 1980, p. 106) of neoteny suggest that embryonic development is the occasion for evolutionary modifications by means of less as well as more development.

Not all the empirical success achieved in this period is explained by "truth in the theories." For example, some of the fossil evidence for linear and parallel evolution turned out not to be robust when new paleontological findings did not conform to the expected pattern (e.g., the evolution of the horse). And for example, Galton's Law of Ancestral Inheritance correctly predicted, for some traits, that there is regression to the mean. The reason for this phenomenon is that some traits are determined by multiple genetic factors whose various combinations result in a normal distribution of the trait. Galton's Law of Ancestral Inheritance makes similar predictions—but without having anything right about the genetics (as we now understand it). The predictive success is, from our point of view, fortuitous.

In some cases we are not yet in a position to assess truth in evolutionary theories. Mivart, Owen, Cope and Hyatt may have been right in thinking that there are internal constraints on the possibilities for evolutionary change (this is now suggested by, e.g., Gould 1993, p. 385). If they were right, this explains the empirical success some of their work in paleontology, e.g., the case of evolutionary senescence of cephalopods. If not, the successes will be explained by the incompleteness of the fossil record at the time (i.e., perhaps we will make new fossil discoveries), or by some truth in orthogenetic theory not yet thought of.

Genetics before the discovery of DNA

The beginning of modern genetics is usually associated with the rise of the Morgan school and the creation of the journal *Genetics* in 1916. This school of genetics, often referred to as "Mendelism," had early and continued success in predicting and explaining the phenotypes produced in breeding experiments in *Drosophila* and other organisms. Phenomena such as genetic linkage, crossover, and eventually genetic mutation were discovered. This success, compounded by the political strength and savvy of Morgan and the opportunities for expansion and innovation in the American academic community, resulted in the creation of a new discipline, a large following in the United States, and knowledge of Mendelism worldwide.

There was not, however, consensus that Mendelian genetics was the correct or complete theory of inheritance. Embryologists, and also European biologists, often rejected classical genetics on theoretical grounds.[11] The research that they did, on cytoplasmic inheritance and the inheritance of acquired characteristics (usually thought to be via cytoplasmic mechanisms), produced some robust results during the interwar and immediate postwar years. These results were not as numerous as those in Mendelian genetics: non-Mendelian inheritance was found in twenty or thirty cases, while Mendelian inheritance was found in around a thousand cases. Many thought that the reason for this is that Mendelian inheritance governs trivial traits such as height and eye color, whereas non-Mendelian inheritance is responsible for basic metabolic and developmental processes, for which variation usually proves fatal. There was experimental support for this idea: many cases of cytoplasmic inheritance involve photosynthesis, respiration and reproduction.

Some examples of non-Mendelian inheritance—either cytoplasmic inheritance, or environmental modifications, or both—discovered during this period are: inheritance of chlorophyll and male sterility in maize, inheritance of male sterility in flax, environmental modifications ("Dauermodifikationen") in *Paramecium* as studied by Jollos, the results of merogeny experiments (nuclei transplanted into different eggs), variegation in many plants, inheritance of mating type, Kappa and antigenic traits in *Paramecium,* pigmentation in guinea pigs, sensitivity to carbon dioxide in *Drosophila* and respiratory deficiency in yeast and

Neurospora (see Sonneborn 1950 and Sapp 1987). Some of these results depended on the development of sophisticated experimental techniques: notably the merogeny experiments, Sonneborn's techniques of controlling breeding in *Paramecium* and Ephrussi's production of petite mutations in yeast.

We know now that there is truth in both kinds of approach (Mendelian and non-Mendelian) that explains the empirical successes of each. Many traits are inherited in Mendelian fashion. Some traits are inherited maternally, through the cytoplasm. Some environmental effects are inherited, at least for a few generations (for example the inheritance of adaptive enzyme formation in yeast). Nuclear-cytoplasmic, nuclear-nuclear and intercellular interactions are all necessary for the regulated expression of inherited traits (so, the nucleus does not rule the cell).

Debate over Continental Drift, 1915–1950

Publication of Wegener's *On the Origin of Continents and Oceans* (1915) kindled widespread debate. It added an alternative to the theories already in use: permanentism and contractionism.[12] At first, the only empirical success of Wegener's drift hypothesis was that it provided an explanation of geophysical, paleontological and paleoclimatological evidence that was hitherto unexplained. This was not enough to be persuasive to most geologists, who thought that drift raised more serious explanatory difficulties than it solved: no plausible mechanism for continental drift through rigid ocean floors had been proposed. Some geologists, who noted that the alternative theories had explanatory difficulties that were just as serious, chose to work with drift.

Although the dramatic successes of mobilism came during the late 1950s and early 1960s, there were definite empirical successes earlier. Alexander Du Toit, regarded by some eminent colleagues as "the world's greatest field geologist" (see Le Grand 1988, p. 82), carried out studies successfully matching geological features in South America, South Africa, India and Australia up until his death in 1949 (see Le Grand 1988, pp. 82–83, and Stewart 1990, p. 39). Some of these studies were early enough to be included in later versions of Wegener's book. Du Toit, and his successor King, worked with an acceptance of mobilism (although not an acceptance of the particular details of Wegener's theory). They looked for the geological similarities that mobilism predicted, and they found them.

During the 1930s, Arthur Wade used drift as a tool to successfully predict the location of oilfields in New Guinea. He assumed, at first as a working hypothesis, that Australia was moving into New Guinea. Later, after finding a fit of the geological features of Western Australia and Antarctica, Wade fully accepted drift (Le Grand 1988, p. 86).

A new technique for measuring absolute geological times, radioactive dating, was developed by Arthur Holmes, and announced in 1913. Holmes went on to theorize that radioactive heating in the earth told against contractionism; indeed, it suggested that there were mantle convection currents that could cause continental drift. Holmes's theory had many of the elements of Hess's later hypothesis of seafloor spreading, which was the background for plate tectonics. Although his theory was not widely embraced, it was widely known, because Holmes was prominent and his *Principles of Physical Geology* (1944) was a popular textbook in Britain. Holmes added radioactive heating to the list of phenomena that were explained by drift.

Finally, the Dutch geologist F. A. Vening-Meinesz developed a technique for measuring gravitational forces at sea. He found, during the 1930s, that gravitational force was much less over oceanic trenches and explained it in terms of descending convection currents.

Stabilists (this includes both permanentists and contractionists) also had some empirical successes. Much fieldwork in the Northern Hemisphere (especially in North America) was done with stabilist assumptions. Some stabilists, most notably Jeffreys, devoted themselves to showing that everything that drift explained could also be explained in a stabilist framework. Jeffreys even argued against Holmes—using different assumptions about the distribution of radioactive materials—that the earth was cooling and contracting.

Wegener's theory of drift was false in many ways, although it was correct about the relative displacement of continents over time. The early successes of drift theory can be explained by this correctness. Interestingly, Le Grand (1988, p. 270) writes, agreeing with Frankel 1979, "It seems that the [Lakatosian] 'hard core' of Drift could be defined only retrospectively and in vague terms: there was no agreed, explicit core belief set down by Wegener which held for all subsequent versions of Drift or for all Drifters." I can hardly think of a better way to state "whig realism" for the case of continental drift.

Fixist views were false, although there was a little truth to them: over relatively short periods of geological time, fixist principles (such as isostasy and uniformitarianism) describe changes in ancient, relatively geologically inactive areas, such as many areas in the Northern Hemisphere. Much work can be done in North American field geology without appeal to tectonic forces. Some fixist empirical successes—such as Jeffreys's explanatory successes—were ex post facto and unremarkable, not in need of explanation.

These cases show that empirical success is not regularly accompanied by either truth (even partial or approximate truth) of a theory or consensus on a theory. It may be possible to tell, in hindsight, what truth there is to past scientific theories, although it is typically not possible to tell around the time that the theories are developed and tested. It is even possible to explain, again in hindsight, significant empirical successes of theories in terms of there being "some truth to the theories." Nevertheless, at the time that theories are developed and tested and for an indefinite period thereafter (until truth about the matter is established—if ever), correct judgments about truth are not available. Any worthwhile methodological advice, therefore, cannot be based on judgments about truth. While it is reasonable to think that there is some truth to, e.g., Freudian psychodynamic theory, genetic algorithms as a model of cognitive functioning, string theory in quantum physics, and even acupuncture, we do not know exactly what in these theories is true and thus cannot use such knowledge to decide how to proceed in further theorizing. At best we work with educated guesses about the theoretical (sub)structures responsible for empirical success.

7 Methodological Import of Whig Realism

The reflections in the previous paragraph could easily be thought to imply that whig realism can offer no methodological advice while traditional realism and antirealism, which do make judgments about truth, have methodological implications. Curiously, exactly the reverse is the case *in practice.*

Traditional realists claim that empirically successful scientific theories are, at core, true. If this is taken seriously, a methodological upshot is that, if truth is a goal, new theorizing should preserve the core claims of the previous theory. No realist whose work I have read actually recommends this (unless, of course, they specify the "core claims" whiggishly, which would make their recommendations methodologically useless). It would be unwise to do so, because numerous scientific advances, from Galileo's heliocentric theory to Harvey's theory of the circulation of the blood to contemporary relativity theory and quantum mechanics have rejected core claims in previous theories. Realists, in practice, tend to make more modest suggestions, such as that new theorizing should make use of empirically successful previous theories in some way. While this is sensible methodology, it derives more from experience (induction on the history of science) than from traditional realism.

Antirealists claim that empirically successful theories are not likely to be true, in whole or in part. If this is taken seriously, a methodological upshot is that new theorizing should be indifferent to previous theorizing. No antirealist whose work I have read actually recommends this. Instead, antirealists reason pragmatically: it is easier, they claim, to work with already existing ideas than to think up new ones (see, e.g., Fine 1986, pp. 119–120). This pragmatic claim is independent of antirealism, so this is not antirealist methodology per se.

Whig realism claims that there is some truth in empirically successful theories. If this is taken seriously, a methodological upshot is that new theories should try building on various portions of previous theories—not necessarily the core claims. Because it is not possible to tell at the time what is true about an empirically successful theory, new efforts will involve trial and error. This is reasonable methodology, in accord with past successful strategies. Thus whig realism is the only version of realism that yields plausible methodology.

Similar results are obtained when considering other methodological consequences of realism, antirealism and whig realism. Another methodological strategy sometimes mentioned is the practice of conjoining successful theories (e.g., Mendelism and Darwinism). Realists expect this strategy to be quickly successful; antirealists expect this strategy to be successful only by chance, and whig realists expect that success is possible, but only after much tinkering (the true portions of each theory, which may not be the central claims, need to be brought together). The whig realist methodology is the most plausible (in the light of historical experience) and realists and antirealists often adjust their recommendations to this methodology.

Yet another methodological strategy, often implicitly assumed, is that of attempting to come to consensus. Realists aim for this, and expect it, because of their conviction that theories can be approximately true. Antirealism itself leads to indifference about consensus, although typically, and curiously, antirealists (e.g., Laudan, Van Fraassen) prefer consensus.[13] According to whig realism, there is frequently progress without consensus when more than one theory has truth in it. Occasionally consensus is desirable, in the rare case that only one theory has all the available truths. This corresponds to the case where only one theory has all the available empirical successes. This will be discussed at length in succeeding chapters. The methodological upshot is a positive attitude towards pluralism, rather than the temporary tolerance for it that is more typical of realist positions.[14] In fields such as formal linguistics and particle physics, long term pluralism is often an explicit methodology of reflective scientists. Whig realism accounts for this pluralism, shows that it is not incompatible with some realism, and suggests that there be more of it.

Arthur Fine's Natural Ontological Attitude (NOA) (Fine 1986), the only current alternative to realism and antirealism, offers no methodological advice either. Although Fine claims that "homely" realist and antirealist attitudes lead to different kinds of theorizing (e.g., Einstein versus Heisenberg[15]) he has no recommendations for which attitude to take in new situations, i.e., he can offer no methodological advice.

Of course, whig realism does not tell the scientist *exactly* what to do when developing theories. It does not generate an algorithm for methodology. Instead, it yields multiple suggestions for trial and error efforts, any one of which has a greater than random possibility of success. Scientific methodology, at its best, is heuristic in character, and requires flexibility and ingenuity to implement. More will be said about this in later chapters.

4

Decision Vectors

If the mind be not engaged by argument . . . , it must be induced by some other principle of equal weight and authority; and that principle will preserve its influence as long as human nature remains the same. What that principle is, may well be worth the pains of enquiry.

Hume, *An Enquiry Concerning Human Understanding*

1 Decision Vectors and Scientific Rationality

Francis Bacon's *New Organon* (1620) captures the epistemic spirit of the seventeenth century scientific revolution. Reason and experience, when uncorrupted by "Idols" of the mind, produce inductive knowledge of the world. Bacon urged (p. 66) that the understanding be "thoroughly freed and cleansed" by a "fixed and solemn determination" to renounce the "Idols"—the various biases and dogmas that prejudice thinking.

Bacon classified the "Idols" into four types. "Idols of the Tribe" are distortions common to all human beings. Bacon claimed "the human understanding is like a false mirror, which, receiving rays irregularly, distorts and discolors the nature of things by mingling its own nature with it." "Idols of the Tribe" include (in more contemporary terminology) overgeneralization, idealization, confirmation bias, salience and motivated reasoning. "Idols of the Cave" result from individual histories and differences. They include attachment to previous education and habits, and individual tastes. "Idols of the Marketplace" are errors produced by the inaccuracies of language, such as vague definitions. Finally, "Idols of the Theater" come from the influence of theological and outmoded philosophical views.

Bacon was convinced that it is both possible and desirable to free the mind of these "Idols" when doing science. Furthermore, those who do science properly, following the inductive method and avoiding the Idols, will agree with one another on scientific matters. "Disagreement . . . shows that . . . the road from the senses to the understanding was not skillfully laid out"; indeed it is a sign of "error" (p. 73).

Bacon's view, or something close to it, has been dominant in science and philosophy of science ever since. It did not even begin during the Scientific Revolution, although it was renewed then, as part of Renaissance thinking generally. The roots are in Greek thought. Plato encouraged dependence on individual reason, free from the influence of either popular opinion or the passions. Aristotle emphasized careful observation and experience.

With the sole exception of David Hume, this view has not been challenged in scientific and analytic philosophy until the later part of the twentieth century. Hume (1748) argued that scientific knowledge depends on human habits of mind that are not chosen for their rational perspicuity but, instead, performed instinctively. Like other animal instincts, they help us physically survive and flourish. Hume thought that reason, even together with observation, is a feeble method of inquiry. He advocated, instead, paying attention to how good mental habits might be cultivated.

Hume, in my estimation, anticipated the best epistemic ideas of the late twentieth century. He is often read as a skeptic, but this reading incorrectly imposes a traditional epistemic framework onto his thinking. Hume was arguing that knowledge requires *more* than reason and experience, not that knowledge is impossible. Many feminist epistemologists and social epistemologists now take this position.

In chapters 2 and 3, I argued that empirical success and truth are primary goals of science. From this, an instrumental account of scientific rationality follows.[1] Scientific practices that lead to empirical success or truth are scientifically rational. Such practices might not be "logical," "clear," "algorithmic" or "objective." It is a contingent matter which kinds of practices are conducive to scientific success. In saying this, I am agreeing with several others who frequently assess reasoning instrumentally, for example Goldman, Kitcher and Thagard. I differ only in my specific account of the goals of scientific inquiry, argued for in chapters

2 and 3. In what follows, I emphasize both empirical success and truth as primary goals of science, where truth is a frequent explanation of the empirical success. Goldman, Kitcher and Thagard talk only of truth and assess decision-making practices for their conduciveness to truth. Sometimes this difference between us matters, sometimes it does not.

Social epistemologists such as Goldman, Kitcher and Hull have already shown that factors such as competitiveness and desire for credit—which are not traditional components of scientific rationality—can be conducive to scientific progress. For example, they consider cases in which competitiveness between scientists motivates them to distribute their research effort over all the reasonable research directions, rather than select the theory that is the most plausible so far. Distribution of research effort is generally conducive to scientific success because it is not possible to tell in advance which theory or theories will be empirically successful or true. Thus, competitiveness is conducive to scientific success. As Kitcher puts it (1993, p. 305), "Particular kinds of social arrangements make good epistemic use of the grubbiest motives."

Feminist epistemologists have argued, furthermore, that ideologies and values of scientists can create and foster directions of research that lead to scientific success. Evelyn Fox Keller (1983), for example, has argued that Barbara McClintock's interactive, respectful attitude towards natural phenomena was crucial for her discovery of genetic transposition.

If motives, values, ideologies and so forth can be conducive to scientific success, they deserve, no less than traditional values of science such as simplicity, fruitfulness, consistency, etc., the status of "scientifically rational." Thus the widespread practice of calling them "biasing factors," which suggests undesirable irrationality, is inappropriately judgmental. To be sure, motivated reasoning and ideology and value-infused reasoning are not invariably or perhaps even regularly conducive to scientific success, so they are not *always* scientifically rational. They are scientifically rational in particular circumstances—perhaps social circumstances or specific scientific domains.

It is time to use a terminology other than "biasing factors" for the social, motivational, cognitive, ideological, etc., factors. I coin the term *decision vectors,* as an epistemically neutral term. Decision vectors are so called because they influence the outcome (direction) of a decision. This influence may or may not be conducive to scientific success, hence the epistemic neutrality of the term. Epistemic evaluation should come after a neutral description of decision-making. The terminology of decision vectors is also deliberately material (physical, like a vector in causal physical theories) rather than abstract (logical), indicating the commitment to naturalistic epistemology of science. Moreover, the terminology emphasizes the fact that decision vectors, whatever their differences in causal origin, operate in the same causal domain.

The term "decision vectors" is also more accurate than terms in widespread use such as "social factors," "external factors" and even "values." Many factors affecting scientific decisions are neither social nor external, nor are they all values of some kind. For example, "cognitive bias" is none of the above. "Motivational bias" is neither social nor value-driven.

There is no shortage of ideas about how the presence of decision vectors can make for scientifically rational choices:

1. Some (e.g., Giere 1988) claim that social, motivational and ideological decision vectors are few and/or weak, and overwhelmed by "rational" factors in successful science. As mentioned in chapter 1, I call this the *Ivory Soap* model of scientific reasoning: the soap is 99.44% pure, or pure enough to be counted as pure. Similarly, reasoning is not significantly contaminated by such (few, weak) decision vectors.
2. Some (e.g., Giere 1988, Thagard 1989) suggest that "cold cognitive" decision vectors such as information processing heuristics are rational, while other decision vectors, notably the motivational ones (so-called "hot cognitive") are not. They also claim that "cold cognitive" factors typically dominate.
3. Some historians of science (e.g., Sapp 1987, Taubes 1993) implicitly suggest that decision vectors *internal* to the scientific process benefit science, while decision vectors *external* to the scientific process do not. So, for example, scientists' motivations to succeed, and power relations within the scientific community, are appropriate whereas external pressures such as religious institutions and the support of non-scientists (e.g., members of Congress) are not.

Approaches (1), (2) and (3) are not ones that I will take. Nor will I adopt any of the next three approaches, although my position has more in common with them.

4. Social epistemologists (e.g., Goldman 1992, Kitcher 1993, Solomon 1992) argue that decision vectors, while not rational at the individual level, tend to be conducive to scientific success at the social level, e.g., by bringing about appropriate distribution of research effort. I call this the *invisible hand* model, after Adam Smith, who argued in another domain (economic theory) that the impulse of self-interest, in a laissez faire economy, will bring about the desirable goal of public welfare.
5. Longino (1990) argues that ideological decision vectors benefit science when the scientific community is structured so as to be, in her sense, "objective." This requires equality of intellectual authority, public forums for criticism, responsiveness to criticism, and shared standards including the standard of empirical adequacy.
6. Finally, social constructivists who claim that scientific rationality is constructed by scientific activity regard all decision vectors as potentially rational (e.g., Collins 1985, Woolgar 1988). For them, "rational" is a term that yields to historical analysis only; they do not themselves use it for normative purposes.

These six different approaches do not of course exhaust the options for an epistemology of decision vectors. In fact, the approach I shall take (*social empiricism*) differs from each of the above. Not to give the punch line away (this will come in later chapters), it shares with (4), (5) and (6) the claim that, when considering decision vectors, scientific rationality is socially emergent. It differs from (4) (5) and (6) as regards the particular conditions for rationality.

Since scientific rationality is assessed instrumentally, it will be possible to decide which normative account of decision vectors is correct. Examination of cases of scientific success will show the conditions under which decision vectors are conducive to scientific success. This will be the task of later chapters.

2 Survey of Decision Vectors

In the remainder of this chapter, I will give some idea of the range of types of decision vectors, and the influence they have on scientific decisions. Different historical, sociological and philosophical narratives have focused on different types of decision vectors. For my purposes, however, it is important to have a balanced and comprehensive account. Bacon's account is almost as comprehensive as any contemporary account. My classification of decision vectors is more inclusive, and organized differently, since my normative purpose is not to eliminate "Idols of the mind," but rather, to recruit and organize decision vectors for scientific success.

Some decision vectors are familiar. For example, pride and conservativeness (which lead to resistance to change belief), radicalism (leads to readiness to change belief), deference to authority (perpetuates the beliefs of the authority) and peer pressure (increases community agreement) are all regarded as "hot cognitive" or "motivational" factors, and are psychological processes within individual scientists dependent on both personality and interpersonal relations. Ideology, funding practices, gender relations and national differences in institutional structure are generally regarded as "social" factors. Some decision vectors operate at the individual level (e.g., pride), some at the group level (e.g., funding practices).

Other decision vectors are either less familiar, or unlikely to be thought of as decision vectors. These include the so-called "cognitive biases" such as salience and availability, decisions made in accordance with theoretical values such as elegance and simplicity, as well as all other "rational" reasons for choice such as consistency and even empirical success. So *the scope of "decision vectors" is wider than the scope of "biasing factors."* Recall the definition of decision vectors given above: *decision vectors influence the outcome (direction) of a decision.* No epistemic judgments need be made in order to describe a factor as a decision vector.

I categorize all decision vectors as either *empirical* or *non-empirical.* This classification will have epistemic importance in the remaining chapters of this book. *Empirical decision vectors are causes of preference for theories with empirical success, either success in general or one success in particular. Non-empirical decision vectors are other reasons or causes for choice.*[2] So, for example, the cognitive salience of a particular set of data for a scientist, perhaps salient because that scientist directly observed it, is an empirical decision vector. The scientist prefers the theory that predicts, confirms or explains those salient data—even though other robust data may be predicted, confirmed or explained by another theory.

Preference for a complex interactive theory is, by comparison, a non-empirical decision vector. The scientist's preference for that theory over, say, a more simple linear theory, is not in itself a preference for a theory with empirical success. Whether or not such a theory is empirically successful is independent of its complexity and non-linearity.[3]

This classification into empirical and non-empirical decision vectors should not be surprising after the discussions in chapters 1 and 2 in which empirical success was found to be a primary goal of science. It is a classification with epistemic purposes that will become clear in the next chapters.

It is instructive to peruse the kinds of decision vectors and their classification into empirical and non-empirical. Here is an extensive, but incomplete list:

Empirical Decision Vectors

Salience of data ("cognitive bias") An empirical decision vector because preference for a theory with salient data is a preference for a theory with *some* data, i.e., a theory with some empirical success.

Availability of data ("cognitive bias") Similarly, preference for a theory with available data is a preference for a theory with some data.

Egocentric bias towards one's own data ("motivational bias") Similarly, preference for a theory supported by one's own data is a preference for a theory with some data.

Preference for a theory which generates novel predictions This is not thought of as a "bias" at all, but it is an empirical decision vector, according to the definition of empirical decision vectors. Sometimes this is not mentioned in a tally of decision vectors, since it is a preference that typically distributes research effort equally over theories with novel empirical success.

Non-empirical Decision Vectors

Ideology ("social" factor) A non-empirical decision vector because it is preference for a theory agreeing with an ideology favored by the scientist(s), independent of any empirical success of the theory.

Pride ("motivational" factor) Non-empirical because it is preference for a theory that enables the scientist to maintain self-esteem and perhaps public esteem, independent of any empirical success of the theory.

Conservativeness ("motivational"/"personality" factor) Non-empirical because this preference for the theory one knows or is already committed to is independent of its empirical success.

Radicalism ("motivational"/"personality" factor) Similarly, non-empirical because this preference for a new, challenging, theory is independent of its empirical success.

Elegance (theoretical value) Preference for an elegant theory is preference for aesthetic features of the theory, which are independent of its empirical success.

Simplicity (theoretical value) Same as for elegance.

Representativeness heuristic ("cognitive bias") This leads to preference for a theory that postulates a particular similarity between two domains based only on some other discovered similarity.[4] Typically (although not always) it is a non-empirical decision vector because similarities are not in general predictive of other similarities and in particular not predictive of similar empirical success. For example, Wegener's claim that continents move through the ocean bed like icebergs drift through water (Wegener 1915, p. 37) was based on the similarity between continents and large icebergs—both large solid masses. However, this similarity does not portend others—continents, for example, cannot move through ocean floors without fragmenting. Empirical success with reasoning about icebergs does not lead, by analogy, to empirical success with reasoning about the movement of continents. In fact, Wegener confessed, "The Newton of drift theory has not yet appeared" (Wegener 1915, p. 167); he admitted that he did not have either theoretical or empirical success in reasoning about the dynamics of the motion of continents.

Competitiveness ("motivational" factor) This is a non-empirical decision vector because the strategic adoption of a theory is independent of the theory's empirical success.

Peer Pressure ("motivational" factor) Also independent of a theory's empirical success.

Deference to Authority ("social"/"political"/"motivational" factor) Likewise independent of a theory's empirical success.

Agreement with Scripture ("social"/"political" factor) As with ideological factors, this is independent of a theory's empirical success.

Note that the basis for classifying decision vectors as empirical or non-empirical is itself empirical, *not* conceptual. It is possible that we will discover, for example, that in some kinds of cases the representativeness heuristic is an empirical decision vector. This is the case when similarities are in fact correlated (typically, when the domains overlap and there can be a causal connection), for then preference for a theory postulating similarities between two domains based on some other discovered similarity is likelier to result in empirical success. For example, Kilmer McCully argued that ordinary atherosclerosis occurs by a similar process to atherosclerosis in children with homocysteinurea. Or, if simple theories could be shown to be empirically more successful than complex theories (either as a matter of logic or by induction), then simplicity would count as an empirical decision vector. I have classified decision vectors according to our current knowledge of their causal connection to empirical success. The classification of most decision vectors is obvious, although still a contingent matter. For example, it is obvious to us that a preference for agreement with scripture will not produce choice of empirically more successful theories.

The list shows that the distinction between empirical and non-empirical decision vectors is *not* identifiable with a distinction between "unbiased" and "biased" decision vectors, "cold" cognitive and motivational ("hot" cognitive) decision vectors, individual and group decision vectors or internal and external decision vectors. Nor does it correspond to the traditional distinction between evidential and theoretical constraints: neither salience nor peer pressure, for example, would be regarded, in traditional epistemology of science, as (rational) constraints on theory choice. Actually, I regard the distinction between empirical and non-empirical decision vectors as the *successor* to the traditional distinction between evidential and theoretical constraints. This will be explained in succeeding chapters.

Considerable multidisciplinary effort has already been expended to identify the decision vectors in scientific choice. Narrative methods, such as those used by historians, are skilled at discerning a variety of decision vectors, especially the motivational, social, political and ideological causes of theory choice. Recently, new techniques have revealed the strength of such previously unnoticed decision vectors as birth order effects, gender ideology, and so-called "cognitive biases." Here is a quick summary of this work, with an emphasis on the "cognitive biases" since they have rarely been discussed in the context of scientific thinking.[5]

Frank Sulloway's *Born to Rebel* (1996) shows that birth order is the single factor most predictive of scientific choice between competing theories in twenty-eight historical scientific revolutions. On average, laterborns are twice as likely as firstborns to choose the more radical scientific theory of two available alternatives. Using a quantitative multivariate analysis, Sulloway examines around 50 factors (such as age, social class, religion and sex) influencing theory choice. Birth order is the best single predictor of theory choice when there are two available alternatives, one more radical or innovative than the other. Birth order produces its effects via the impact of birth order on general personality traits. Eldest children are thought to identify more with parental authority, and thus be more conservative.[6] The effect of birth order is not general or universal: when competing theories do not differ in ideological radicalism, birth order effects are much smaller.

Feminist scholars have documented both manifest and subtle effects of gender ideology on theory choice. When there are sex differences in nature, such as the differences between gametes from male and female organisms, or the brains of male and female persons, gender differences such as differences in passivity and activity and rationality and emotionality are often projected onto nature. Sometimes this is correct, and sometimes mistaken. Longino (1994) calls these "specific" assumptions. There are also subtler, or, as Longino calls them, "global" assumptions such as that natural process are linear and governed by single causes, rather than interactive and multiply caused. Longino (1990) as well as Keller (1985) and others associate linear thinking with masculinist, hierarchical ideology, and interactive thinking with feminist, democratic ideology.

Experimental work by Tversky, Kahneman, Nisbett, Ross and others on the psychology of belief change shows that human decision-making is frequently guided by a few general heuristics such as salience, representativeness and availability (see, for example, Nisbett and Ross 1980 and Kahneman, Slovic and Tversky 1982).[7] Subjects weight salient and available information more heavily in their decision making than normative models from probability and confirmation theory suggest that they should. Factors making data salient include concreteness (the detail, even irrelevant detail, with which things are described or actually experienced), proximity, emotional interest and perceptual biases. Factors influencing availability include biases in exposure and attention to data, biases in memory retrieval, and salience. The heuristics of salience and availability are also responsible for the phenomenon of belief perseverance (sometimes called "confirmation bias"). Individuals hold onto beliefs at least partly[8] because memory retrieval and perceptual attention operate in such a way that the evidence for beliefs already in mind is made more available and salient, and therefore weighted more heavily, than the evidence against them. The representativeness heuristic underlies, among other things, causal reasoning. Subjects assume that similar events have similar causes, and reason informally, without considering base rates or the scientific relevance of the similarities found.

"Cognitive bias" has particularly interested me (starting with Solomon 1992) because the heuristics of salience, availability and representativeness mostly involve "cold" cognition, that is, cognition free of emotional and motivational influence. This supports my claim that the demarcation between "cold" cognition and the passions is not to be identified with the epistemic distinction between rational and irrational decisions. The heuristics of salience, availability and representativeness have no *more* claim to rationality than "hot" cognitive processes such as wishful thinking, pride and competitiveness.

Here are some examples in which these cognitive heuristics played a crucial role in scientific work. William Bateson, a founder of the mutation theory, found very little correlation between environmental differences and species variation in lakes in central Asia (Bowler 1983, pp. 189–90). In other words, he found little evidence of adaptation. These direct observations were responsible for Bateson's view that natural selection, as well as Lamarckism, was false. Bateson did not take Darwin's observations of finch adaptations in Galapagos as seriously as he took his own observations, even though they were widely accepted. Cognitive salience and availability, as well as perhaps some egocentric motivational bias, explain this.

J. Tuzo Wilson, a Canadian geophysicist who ultimately became a leader of the plate tectonics revolution, was a firm contractionist early in his career. According to Le Grand (1988, p. 190)), Wilson's loyalty to contractionism began during his postgraduate education at Cambridge, where he was influenced by Jeffreys's contractionism (the decision vector here is influence of authority). This conviction was deepened during Wilson's fieldwork in the Canadian Shield. As Le Grand (1988, p. 190) reports, " 'nothing could appear more rigid or unchanging' than the region in which he had a strong interest and expertise (Wilson 1966a)." Evidently, salience and availability of data supporting contractionism were also decision vectors influencing Wilson.

Wegener hypothesized that the mechanism for drift of continents is similar to the mechanism for drift of large polar icebergs (Wegener [1915] 1966, p. 37). This is an example of the use of the representativeness heuristic and also the cognitive salience and availability of Wegener's 1906 observations of icebergs in Greenland. The use of the representativeness heuristic is not a piece of convincing causal reasoning—the situations of icebergs and continents are not shown to be sufficiently similar to support the conclusions drawn. Many rejected drift, claiming that movement of continental crust through oceanic crust without significant distortion is physically impossible.

Additional examples of all types of decision vectors will be given in the next two chapters. Recall that a decision vector is *anything* that influences the outcome of a decision. This includes such traditionally "rational" decision vectors as "preference for theories with empirical success" and "preference for coherent theories." I do not always mention these decision vectors when—as is sometimes the case—they play no role in the distribution of research effort in a case under discussion (for example, when simple theories are generally favored and both theories are perceived as simple).

The number of types of decision vectors is probably between 50 and 100. Decision vectors differ in magnitude from one another and from context to context, and some interact with one another (e.g., Sulloway has shown that there is an interaction between birth order and gender). Any accurate descriptive account of scientific change has to handle this complexity.

Moreover, any useful normative account, in a naturalistic philosophy of science, is dependent on a good descriptive account. A normative account evaluates decision vectors for their conduciveness to scientific success. Normative recommendations for, e.g., adding or removing decision vectors must be realistic.

Some philosophers sidestep this complexity. Either they consider only a few decision vectors (e.g., Kitcher's 1993 and Goldman's 1992 analyses of the effect of competitiveness) or they try to classify decision vectors, generally, as rational and non-rational (e.g., Giere's 1988 and Thagard's 1989 classification of "cold" cognitive mechanisms as rational). These strategies are deficient. Idealized models that consider only a few decision vectors fail to model actual decisions, as I shall show in the next two chapters. And "cold" cognition has no more of a claim to "rationality" than does "hot" cognition, as discussed above.

Sulloway (1996) handles the complexity of factors in theory choice by using a multivariate model. This is a statistical method that, together with some causal knowledge of the situation, is capable of detecting and predicting the contribution of each relevant factor to an individual's decision.

In my analyses, I will use a simple multivariate model—much simpler than Sulloway's—to investigate the effect of decision vectors not on individual scientists' choices, but on the *aggregate* of these individual choices in particular episodes of scientific change. I am less interested in predicting the choice of individuals[9] than in seeing how decision vectors are involved in the aggregated decisions of a community. The standpoint for instrumental evaluation of the effects of decision vectors is thus social. The model, while simple, is adequate for current purposes.

Scientific rationality—conduciveness to scientific success—is not an *intrinsic* property of most decision vectors. As will become evident, the same decision vectors can be conducive to scientific success in some situations and not others. Only the empirical decision vectors are always conducive to some scientific success, and even then, they do not typically maximize attainable empirical success. For all decision vectors, context, in particular the context of all decision vectors operating in a community, is crucial for an assessment of their normativity.

5

Dissent

If all pulled in one direction, the world would keel over.
Yiddish proverb

1 Initial Reflections

Scientists frequently disagree with one another. The disagreement can be short lived or permanent, localized or extensive. The traditional view of scientific rationality is that such disagreement is, and should be, brief and rationally resolvable. Bacon claimed that disagreement is a sign of error (1960 [1620] p. 73). Much more recently, Philip Kitcher writes that "in a community of clear-headed scientists, devoted to espousing epistemically virtuous individual practices, we may expect cognitive uniformity" (Kitcher 1993, p. 344).[1]

Scientists' disagreements are, however, often profound (not due to simple correctable error) and long lasting. As already seen in chapters 2 and 3, this situation is not automatically cause for normative concern. Scientists who disagree with one another can each pursue and achieve empirical successes. Also, there can be truth in each of their theories, in the sense discussed in chapter 3.

Indeed, in the last decade, there has been interest in the distribution of research effort (sometimes called the division of cognitive labor, see, e.g., Goldman 1992, Kitcher 1993 and Solomon 1992). Scientists often have the opportunity to explore a number of potentially successful directions of research, and it is most efficient to accomplish this when different individuals, or groups, pursue different theories or strategies at the same time.

This does not *require* dissent, because scientists can agree about the likely success of the different directions of research while choosing (for a variety of motives or reasons, such as those mentioned by Goldman and Kitcher) to pursue different theories. Dissent is just one kind of cause—perhaps the most common cause[2]—of distribution of research effort. The idea of distribution of research effort is significant because it necessitates a move to social epistemology. The community of scientists is evaluated for how appropriately it distributes research effort.

Some recent accounts of scientific disagreement defend a fairly traditional view of rational cognitive uniformity. Thagard, for example, emphasizes the disagreements that result from delays in communication of relevant information, and writes, "Individual scientists each start with a different stock of personal knowledge and experience so even if they have exactly the same rational processes we should expect them to have knowledge bases that do not completely converge because of the impossibility of perfect communication (Thagard 1993, pp. 66–67). And Giere claims similarly that scientists all have the same basic, and rational, reasoning processes while they may apply those reasoning processes in different knowledge bases (Giere 1988, pp. 174–6). Or, for example, subjective Bayesians claim that scientists initially come to different decisions because they assign different prior probabilities to competing theories (e.g., Howson and Urbach 1989). Scientists' decisions often converge after sufficient exposure to relevant additional evidence, a process known as "washing out of the priors." Each of these positions differs on the precise constitution of individual rational processes (they hold, respectively, that rationality is shown in assessments of explanatory coherence, that "cold cognitive" processes are rational and that Bayesian methods are rational). Each of them, however, also takes a social perspective to evaluate the distribution of cognitive effort. Dissent, for these philosophers, is both fortuitous and temporary, expected to resolve in consensus when information is effectively communicated and sufficient evidence is gathered.

Laudan, and Kuhn in later writings, break more strongly with tradition when they argue that scientists' disagreements are due to their different methodologies. (According to tradition, there is only one scientific method.) For example, Laudan, with Rachel Laudan (1989) claims that the early (before about 1960) espousers of drift especially valued general explanatory power, and the later espousers of drift especially valued novel predictions and independent testability. Drift, the Laudans' claim, only had explanatory power before the mid-1960s; not until the magnetic reversal patterns discovered by Vine and Matthews did it have independent predictive success. On this view, consensus occurs when a theory is superior according to all methodologies.[3] Dissent results in a useful division of labor, so that each plausible theory will be pursued. Kuhn, in "Objectivity, Value Judgment and Theory Choice" (1977) has a similar view that the generally shared criteria of choice (predictive accuracy, consistency with current theories, broadness of scope, simplicity, fruitfulness of new research findings) do not come with a shared algorithm for weighing them, and that difference in individual judgments produces the necessary division of labor (pp. 331–332). He takes himself to be showing "the meaning rather than the limits of objectivity" (p. 338).[4] In both of these views, rationality is thought of as a property of the decision processes of individual scientists, and motivational and social factors are not part of scientific rationality. Consensus takes a while to achieve and is dependent on one theory becoming superior according to all methodologies.

These contemporary views have much in common with one another, despite their disagreements with regard to the actual causes of dissent. They are all rather laissez-faire about the distribution of cognitive effort, either assuming or arguing (using abstract models) that the causes of dissent distribute cognitive effort appropriately, with more resources to the more promising theories.[5] This is the "invisible hand of reason" view described in the previous chapter. And they all view consensus on one theory as the end point—whether temporary or permanent—of a period of inquiry, ending division of effort and ending dissent. A social epistemic perspective, while it might be useful during periods of uncertainty and/ or dissent, is abandoned when discussing consensus.[6]

I have a different view about dissent. I find cases where research effort is poorly distributed as well as cases where it is well distributed. Obviously, for example, it would be undesirable for an entire scientific community to pursue just one of a variety of promising theories or to ignore a theory with empirical success. This points to the need for a normative epistemology of dissent to analyze distribution of effort and improve it where necessary. I give this in this chapter. In chapters 6 and 7, I argue that a social epistemic perspective is as useful for assessing consensus as it is for dissent.

Decision vectors are powerful and plentiful. A proper account of the decision vectors operating in particular cases shows the causes of disagreement among scientists. As the cases below will show, disagreement is not attributable to a few traditionally "rational" factors such as difference in prior knowledge or difference in methodology. Nor is it attributable to a few "reasonable" motivational factors such as the desire for credit or reward. Disagreement is caused by multiple decision vectors, some "rational," some "reasonable," some decidedly "nonrational," "unreasonable" or even "irrational" by traditional standards. Any account of the effect of decision vectors needs to be multivariate. I will use a simple model, called an "improper linear model," to begin the analysis. This will be explained in the context of the first and lengthiest case study.

2 Evolutionary Biology after the Publication of *Origin:* An Example of Good Distribution of Research Effort

As already discussed in chapters 1 and 2, there was much dissent in evolutionary biology during the late nineteenth and early twentieth century. Four theories of evolutionary change were in use, in various combinations: natural selection, saltative evolutionary change, neo-Lamarckism and orthogenesis. In this chapter, the focus will be on the reasons and causes of that dissent: on the decision vectors influencing individual scientists and groups of scientists. I will tell the historical tale of dissent again, but this time discuss what was going on *besides* production of empirically successful theories.

Origin sets out a theory of evolution in which species change is gradual, adaptive and non-progressive, and produced by physical mechanisms that operate according to chance. *Origin* was radical for its time—scarcely anyone agreed with Darwin that natural selection is a primary mechanism of species change. The many biologists who called themselves "Darwinian" often meant only that they believed in the natural transmutation of species. It is inappropriately whiggish, therefore, to analyze this case using a classificatory framework that considers only agreement with Darwin's theory of natural selection, the precursor to our present dominant theory of evolutionary change. Such a classificatory framework arranges beliefs about transmutation of species in a one-dimensional way, from belief in natural selection as the main mechanism of evolutionary change, to belief that natural selection is a possible mechanism of evolutionary change, to belief that natural selection does not account for evolutionary change. The variety of other theories is not acknowledged in this framework; indeed, Lamarckism is classified together with Special Creationism, since neither allows that natural selection accounts for evolutionary change. Kitcher (1993), Sulloway (1996) and Thagard (1992) use such whiggish analyses. Even Bowler (1988) classifies biologists into "Darwinians," "pseudo-Darwinians," and "anti-Darwinians," keeping Darwin as the single frame of reference, which is historically inaccurate. (Bowler does, however, go on to appreciate the complexity of the different positions.) I do not classify biologists, but I classify their theories—some biologists adopted more than one, e.g., saltationism together with natural selection—into the four theories just mentioned.

I will first describe the decision vectors affecting Darwin and natural selection in particular and then look at the rest of the community, and other prominent biologists, in the late nineteenth century. The decision vectors are in no particular order. At the end of each description of a decision vector, for future reference, I state whether it is an empirical or non-empirical decision vector, and which theory or theories the decision vector favors (with a +) or disfavors (with a −).

a. Lyell's work was used for making an *analogical inference* between gradual, continuous, non-progressive geological change and gradual, continuous, non-progressive species change. In both domains, Darwin claimed, the mechanisms of change in the past are the same as the mechanisms for change in the present, so that all mechanisms for change can be studied by looking at present processes. This analogy is not supported by any evidence that geophysical change and biological change are similar; it is an example of reasoning driven by the representativeness heuristic[7] where two domains which each have long histories of change are said to change in similar, i.e., gradual and non-progressive ways. (Non-empirical decision vector; natural selection +)

b. Darwin's plentiful observations of the results of horticulture and stockbreeding gave him evidence that species can undergo gradual and significant change when there are selective pressures to do so. He focused on this salient data, rather than on the negative (less salient) fact that there was never an example of transmutation of species by means of artificial selection. (Empirical decision vector; natural selection +)

c. Darwin extended Malthus's model of population growth and competition for survival among humans to the whole of nature, reasoning by analogy. (Non-empirical decision vector; natural selection +)

d. During the five years on the *Beagle,* Darwin was exposed to, and reacted against the conservative Christianity of his captain, Fitzroy. Darwin was frustrated at being unable (because of the unquestioned authority of a captain at sea) to freely express his dissent. As Gould has said (1977, p. 33), "Fitzroy may have been more important than finches, at least for inspiring the materialistic and antitheistic tone of Darwin's philosophy and evolutionary theory." (Non-empirical decision vector; natural selection +)

Further decision vectors apply as much to Darwin's followers—who may have believed in natural selection (e.g., Hooker), in environmentally induced adaptation (e.g., Spencer) or in saltative evolutionary change (e.g., Huxley)—as to Darwin himself.

e. Sulloway (1996) has persuasively argued that birth order is the most important single predictor of radical scientific choice. Darwin and most self-described Darwinists (those who believed in the transmutation of species by any of a variety of natural mechanisms) were later-borns. Birth order is hypothesized to be an important determinant of the personality trait "openness." (Non-empirical decision vector; natural selection +, neo-Lamarckism +, saltationism +)

f. Darwin had good public relations, at least in the United Kingdom. Darwin himself was skillful and deliberately non-threatening in the presentation of his ideas. He claimed that he was following the accepted canons of scientific method and set out his arguments to show this. He waited to write about evolution until he had established his reputation as a naturalist. He insisted that his primary claim was the transmutation of species, and that natural selection was only a secondary claim (Young

1985, p. 109). And his use of the metaphor "selection" permitted others to read a teleological, developmental, progressive narrative—more palatable than blind materialism—into his theories, as Young (1985) has argued. He cultivated prominent and skillful supporters, especially Hooker, Wallace and Huxley, who argued publicly on his behalf. In Britain, at any rate, Darwinism had much better public relations than other theories on the origin of species. (Non-empirical decision vector; natural selection +, neo-Lamarckism +, saltationism +).

g. From (f) it should be noted that many who called themselves "Darwinists" misunderstood the theory of natural selection, and in fact believed that evolution is progressive and dependent on individual effort, rather than on blind materialistic forces. Such progressivist views were sometimes appealing because they cohered with widely held views about social change, particularly the Victorian ideology of capitalism (e.g., Huxley, Haeckel, Butler, Spencer). More people became Darwinists because of this misunderstanding of Darwin's views, which Darwin was less anxious to correct than was, e.g., Weismann later on. (Non-empirical decision vector; natural selection +, neo-Lamarckism +, saltationism +)

h. Darwinism was especially appealing to naturalists (e.g., Darwin himself, Edward Bagnall Poulton, Wallace, Hooker and the Harvard botanist Asa Gray), for whom adaptive traits were often salient from field observations. (In contrast, as I will discuss below, paleontologists and developmental biologists tended not to be Darwinists.) Neo-Lamarckian adaptive mechanisms, for the same reason, were also appealing to naturalists. (Empirical decision vector; natural selection +, neo-Lamarckism +)

It took a while for the serious alternatives to Darwinism to be produced, but by the late nineteenth century, neo-Lamarckism, orthogenesis and the mutation theory (based on the idea of saltative evolutionary change) had widespread appeal. Decision vectors were involved in the opposition to Darwinism:

i. It is popularly thought that conservative religious views about creation were the opposition to Darwin. In fact, within the community of naturalists and biologists, it was widely accepted by the 1860s that species evolve, and by the 1880s even theistic evolutionism lost its popularity. Religiosity played a more subtle role, however, in those who had a preference for non-materialistic theories of evolutionary change. There were two candidates for such theories: idealist forms of orthogenesis,[8] and voluntarist versions of Lamarckism. Agassiz, Owen and Mivart were prominent biologists whose religiosity led them to orthogenesis. Samuel Butler thought of growth and differentiation as a teleological process in which the individual's desires and habits control the direction of evolution. (Non-empirical decision vector; neo-Lamarckism +, orthogenesis +)

j. Progressivists who were not necessarily religious objected to the radical materialism in Darwinism. Natural selection, accurately understood, is the view that species evolve as a result of random variation and survival of the fittest, a process that leaves many casualties and extinctions along the way. Many found this theory too painful to accept. Some of them declared their opposition to Darwinism (e.g., Morgan, Bateson, Mivart and Galton) and claimed that species change occurs through saltations or through Lamarckian processes. Others (mentioned in (f) and (g) above, e.g., Hooker, Spencer, Haeckel), while nominally Darwinists, made extensive appeal to Lamarckian, orthogenetic and saltationist ideas. Those who called themselves Darwinists were not always closer to "true" Darwinism (i.e., natural selection) than those who explicitly opposed Darwin. (Non-empirical decision vectors, neo-Lamarckism +, saltationism +, orthogenesis +)

k. The political ineptness of many of the anti-Darwinians, especially in Britain, decreased the appeal of their theories. Owen and Mivart, two leading anti-Darwinians who believed in orthogenesis, had little impact because of their difficult personalities. (Non-empirical decision vector; orthogenesis −)

l. Paleontological evidence seemed to support the view that species change is discontinuous, usually progressive and sometimes non-adaptive. Cope and Hyatt, for example, each made particular discoveries so salient to them (because they were their own discoveries) that they rested general orthogenetic theories on them. Cope's work on the evolution of the horse through distinct stages (intermediate fossil forms were not found) and Hyatt's work on the evolution of a cephalopod line through to extinction (interpreted with analogy to the life of an organism) carried less weight for Darwinians, who chose to view the discoveries as atypical or incomplete, and for naturalists, for whom living specimens were more salient. It has turned out Cope's observational conclusions are false; intermediate fossil forms in the evolution of the horse have been discovered. Thus, the influence of Cope's observations is classified as nonempirical, due to salience of something other than evidence. (Empirical decision vectors, orthogenesis + and saltationism +; non-empirical decision vectors, orthogenesis + and saltationism +)

m. Morphologists, who often made use of paleontological evidence, but also evidence from living organisms, also tended to be opposed to Darwinism, and in favor of orthogenetic theories of an idealist kind. Eimer's work on coloration in butterflies, which showed non-adaptive and discontinuous change, is an example of this. Morphologists were influenced by evidence that was salient and available to them. (Empirical decision vector; saltation +, orthogenesis +)

n. Idealist philosophy, which claimed an underlying Platonic order to nature, was popular and supported orthogenesis. Agassiz, Cope, Hyatt and Eimer were all metaphysical idealists (see Bowler 1983, p. 147). (Non-empirical decision vector; orthogenesis +)

o. The evolution of species was often looked at through analogy to the growth of an individual organism. Indeed, the word "evolution" was Originally used to describe the growth of an embryo, and it was rarely used by Darwin, who wrote of "transmutation of species." The influential recapitulation theory of Haeckel, for example, according to which "ontogeny recapitulates phylogeny," lent support to progressivist and orthogenetic theories of evolution, rather than to theories of random variation. A further analogy, between inheritance and memory, yielding the idea of species memory as shown in embryological development, was also employed. (Non-empirical decision vector; Lamarckism +, orthogenesis +)

p. Birth order, as in (e) above. Sulloway (1996) found that not only Darwin and most Darwinists are later borns, but also that most anti-Darwinians, who tended to be orthogeneticists, were firstborns or singletons (no siblings). Firstborns tend to be more conservative—less open to new ideas, and more traditional—than laterborns. (Non-empirical decision vector; orthogenesis +)

q. Some countries were organized so as to bring about widespread rejection of Darwinism. For example, France was generally anti-evolutionary because of the influence of the Catholic church, scientifically rather isolated, and chauvinistic (so, if sympathetic to any evolutionary theory, it was to that of their countryman Lamarck). And, for example, Agassiz, who was staunchly anti-Darwinian, believing in orthogenesis, had great influence in the United States. His research was well funded—much better than, for example, the Darwinist Asa Gray's research—and he had powerful students. (Non-empirical decision vectors; neo-Lamarckism +, Orthogenesis +)

r. Galton compared transmutation of species to the motions of a multifaceted stone, or polygon. This analogy was influential in communicating and making plausible a discontinuous (saltative) theory of species change. The idea is that minor variations within species—either inherited or environmentally induced—correspond to the stable equilibrium of a slightly wobbling polygon, and never add up to transmutation of species, which corresponds to the rolling of a polygon onto another face. Some other process, therefore, is responsible for species change. Mivart, for example, explicitly took up this analogy for his saltative theory. (Non-empirical decision vector; saltation +)

s. Weismann, who is known for the "germ plasm" theory (an antecedent of the later chromosome theory) of inheritance, insisted that natural selection is a sufficient mechanism for evolution. He was more Darwinian (in the narrow sense) than Darwin! His insistence that natural selection was the single mechanism of evolutionary change did not convince others. It only polarized the debate and focussed attention on the largely speculative nature of natural selection. His observation that cutting off tails of mice did not result in inherited taillessness did not undermine neo-Lamarckism.[9] At the end of the nineteenth century, those (e.g., Spencer) who had included natural selection in the list of possible mechanisms for evolutionary change now rejected it. (Non-empirical decision vector; natural selection −).

This has been a long case study, and it is now time to reflect on it. I have listed numerous decision vectors, some empirical, some non-empirical. The number, character, strength and pervasiveness of these decision vectors tells against the possibility of an "Ivory soap" account of the debate in which "non-rational" factors play no significant role. Moreover, the number of types of decision vector means that models that focus on one or two types of decision vector (such as Kitcher's, which includes the desire for credit and the influence of authority) do not go far enough. How, then, can the role of decision vectors be normatively assessed?

The case is so complex that it is difficult enough to know how to tell the *descriptive* story, never mind do a *normative* assessment. I have listed the decision vectors, but not determined their effects precisely, i.e., quantitatively. Which had more effect, the influence of Lyell's gradualism or the influence of progressivist Victorian capitalism (for example)? Obviously, we are not in a position, with current data and historiographic tools, to answer this question. It is not surprising that different narrative histories of the Darwinian revolution foreground different decision vectors. Narratives often simplify. Oldroyd (1984) catalogues the many different accounts of the Origin of *Origin,* which emphasize different influences on Darwin's ideas. Accounts of the reception of *Origin* also emphasize different factors: Young (1985), for example, features the role of Victorian culture in the positive reception of *Origin;* Bowler (1983) emphasizes the negative conservative reaction to *Origin.* The virtue of my own account, which lists decision vectors, is that it holds off on the narrative theorization, is more eclectic in approaches to discovering the various decision vectors, and in the process includes more of the historical data.

A complex situation calls for a multivariate analysis. Multivariate analyses are now standard methodology for investigating complex phenomena in epidemiology, economics and the social sciences. These analyses employ statistical methods to discern correlations between several variables and a phenomenon under investigation that is dependent on them. Also, interaction effects between the variables can be detected and measured. The analysis is quantitative. When multivariate analysis is supplemented with causal hypotheses, which account for the discovered correlations, the result is a complete dynamical account of change. Causal hypotheses can come from qualitative studies in which likely causes are identified in particular case studies, or they can come from interventions and observations in which the values of variables are changed so that their relations of dependence and independence are likely to show.

Multivariate models of scientific change have rarely been offered in the science studies literature. The most substantial exception is the recent work of Frank Sulloway (1996), which investigates the correlation of birth order and other variables with theory choice, and also examines these variables for interaction effects. Sulloway includes over forty psychological and social variables for around thirty scientific controversies. However, it is in the nature of this analysis that decision vectors particular to one scientific controversy (such as, for example, the influence of Victorian culture on Victorian science) are not included. Because they are not factors in every controversy (or at least several controversies), their contribution cannot be calculated by the method of multivariate analysis. Moreover, Sulloway does not include "cold cognitive" factors such as reasoning by analogy (use of the representativeness heuristic) and the role of availability and salience. Much more data would be needed than is available in order to include them.

The main reason that a multivariate analysis such as Sulloway's is not applicable here is that it is descriptive, not normative. I am trying to assess the contribution of decision vectors to scientific success, whereas Sulloway (and others who offer descriptive accounts) are only aiming to assess the contribution of (some) decision vectors to the actual historical outcome, which might be successful or not.

For assessing the contribution of decision vectors to scientific success, what matters is the *distribution* of the decision vectors. If decision vectors overwhelmingly act in favor of one theory, and there is more than one theory with empirical success, then the distribution of decision vectors is clearly not conducive to scientific success because some viable theories are not being pursued at all. What counts as a desirable distribution of decision vectors? Bear in mind that there are two kinds of decision vectors, empirical and non-empirical, and that decision vectors differ in magnitude. Also, the goal is not equal distribution of research effort, since some theories—such as those with more empirical success—may deserve more resources than others. The goal is an equitable distribution of research effort, perhaps a distribution that is proportional to the empirical success.

As argued in chapters 2 and 3, empirical success is a primary goal of science. Empirical decision vectors, because they are associated with empirical success, influence scientists to work on worthwhile theories. Non-empirical decision vectors, on the other hand, do not select for or against empirical success, or any other primary goal of science (such as truth).[10] If empirical decision vectors are distributed *equitably,* i.e., in proportion to the empirical success of the various theories under consideration, and non-empirical decision vectors are distributed *equally,* then, overall, the distribution of decision vectors will be equitable. This reasoning is based on the reflections of chapters 2 and 3, and will be given more credence through its usefulness in application to case studies.

Since decision vectors differ in magnitude, and, with present historical techniques, these magnitudes are not exactly determinable, a precise assessment of distribution is not possible. It is not clear that it is even desirable to quantify the magnitude of decision vectors, especially since the amount of empirical success is not currently quantifiable either (see chapter 2). Reasonable estimates are, however, possible and desirable.

The simplest kind of multivariate model is the "improper linear model." In this analysis, variables are not assigned their actual magnitudes nor are their interactions with one another considered. It is simply noted whether the effects are positive (+) or negative (−), and then the effects are summed additively. As Robyn Dawes (1988, chapter 10) has shown, use of an improper linear model is superior to informal (sometimes called "global," or "intuitive" or "seat of the pants") assessments. For example, Dawes has shown that improper linear models outperform the "experts," who make informal judgments in three cases: diagnosis of neurosis versus psychosis from scores on the MMPI, prediction of grade point averages from 10 variables assessing academic aptitude and personality, and prediction of graduate student performance based on GRE, undergraduate GPA and the selectivity of their undergraduate institutions.

Thus we may reasonably expect that an improper linear model will be superior to the usual informal assessments of the effects of decision vectors made by narrative historians, scientists and philosophers. Here is an improper linear analysis of the case of late nineteenth century evolutionary biology, taken directly from the assessments of each decision vector (see parenthesis at the end of each discussion):

Empirical Decision Vectors

Natural Selection	+2
Neo-Lamarckism	+1
Orthogenesis	+2
Saltationism	+2

Non-empirical Decision Vectors

Natural Selection	+5
Neo-Lamarckism	+7
Orthogenesis	+6
Saltationism	+6

Each of the four theories had some empirical success. During the late nineteenth century, no one theory had strikingly more empirical success than any other. The distribution of empirical decision vectors was approximately equitable, therefore. The non-empirical decision vectors were approximately equally distributed. Thus, distribution of research effort was more or less appropriate; no theory got much more or less than its share of attention.

Popular thought, as well as some historians (e.g., Bowler 1983), has suggested that Darwinism was eclipsed by an unscientific conservative backlash that only slowed the eventual triumph of Darwinism. This analysis might result from whiggish history—from characterizing the historical options as either for or against Darwinism, construed narrowly as the theory of natural selection—in which case an improper linear analysis of the decision vectors would result in:

Non-empirical Decision Vectors

Darwinism	+5
Against Darwinism	+18

Empirical Decision Vectors

Darwinism	+2
Against Darwinism	+5

The empirical decision vectors are distributed equitably, but the non-empirical decision vectors are not distributed equally. This can be interpreted as an overall skew against Darwinism, and thus a normatively inappropriate "eclipse of Darwin." I hope I have shown that this whiggish analysis is itself normatively inappropriate, because it does not describe actual choices accurately. Late nineteenth century evolutionary biologists had four directions of research to select from, not two.

Before moving on to the next case study, I want to complete the analysis with some qualifications and clarifications. Some might object that I have not (or may have not) included all the decision vectors in this historical account. A few extra decision vectors here or there would not change the conclusion—that the decision vectors are reasonably well distributed—but a large number of extra decision vectors could result in a skewed distribution. I am open to such criticisms. I have attempted, in this chapter, to draw on a range of respected narrative historical accounts, in order to represent, fairly, our current state of knowledge about decision vectors. I am not attempting to make an original historical contribution, simply to do a new kind of normative analysis. If historical understanding changes, normative judgments will change along with that understanding.

There may also be questions about the period of time over which decision vectors should be assessed for distribution. Any period of time *can* be chosen. Above, I chose a period of about forty or fifty years after the publication of *Origin.* I did this in order to answer the question, "How well was cognitive labor distributed during the "eclipse of Darwin" period?" If the interest was in the period 1860–1870, considerably fewer decision vectors would be considered, and the normative assessment might come to a different conclusion.

This analysis may also create the impression that decision vectors typically "balance out,"[11] leaving rational, or "scientific" factors, or even quasi-rational (because associated with empirical success) factors such as the empirical decision vectors to do their work. Like Adam Smith's invisible hand in which myriad actions of self-interest result in a good general distribution of wealth, the non-empirical decision vectors balance out in the end, leading to a good distribution of cognitive effort. I call this the view that there is an invisible hand of reason. It is not, however, my view.

There is no a priori reason to think that non-empirical decision vectors will always be roughly equally distributed between available theories. And one case study showing roughly equal distribution of non-empirical decision vectors—late nineteenth century evolutionary biology—is not much of an a posteriori reason. If non-empirical decision vectors are independent of one another then, statistically, if there were many of them, we would expect a reasonably good distribution. But we do not know how independent of one another they are, and, moreover, the number of decision vectors is not generally large enough that their distribution obeys the laws of large numbers. To show how decision vectors can have other outcomes, I present some more case studies.

The situation with empirical decision vectors is different. Typically, there are empirical decision vectors associated with each empirical success, since empirical successes are produced by individuals with special relations of salience and availability (because of concreteness) to the data. This leads to a distribution of research effort over all the theories with empirical success, and more over those with more empirical success. Occasionally, however, extra empirical decision vectors are associated with an empirical success because of some special feature of that success, e.g., it is especially vivid.[12] Thus, while it is less likely that there is an inequitable distribution of empirical decision vectors, this can still happen.

A final point. Some might object that it is inappropriate to make normative judgments, which entail historical counterfactual claims of the sort, "if decision vectors had been better distributed, scientific ideas would have progressed further or faster or better." This objection has two parts. First, is it not whiggish to think one knows better than the scientists at the time? I have addressed this objection by being careful to present the choices faced by scientists non-whiggishly. I give the options that presented at the time. Second, historical events are so complex, and there are so many interacting causal factors, that historical counterfactual claims may be no more than fantasies. Who knows that would have happened if the historical antecedent conditions had been different this way or that? My answer to this concern is to suggest where the burden of proof lies. If we cannot make reasonable historical counterfactual claims then we implicitly condone all past action and remain agnostic on current choices ("Who knows what the effects may be of this or that action?

Everything is so complex"). I find this position both implausible and irresponsible. Of course, there is *fallibility* in historical counterfactual claims, but we are far from ignorant. Normatively speaking, and speaking only with respect to science (although similar points might be made with respect to other historical change, such as political change), I think it is far better to run the risk of getting some normative judgments wrong than it is to shy away from making any at all. Refusal to take a stand on past decisions is also invariably associated with refusal to get involved in current controversies, a stand that is both irresponsible (at least, for matters of public interest, such as health and environment) and difficult to consistently maintain.

3 Genetics before the Discovery of DNA: An Example of Less Good Distribution of Research Effort

Genetics developed differently in different countries. In the USA and the UK, Mendelism dominated by about 1920. In Europe, there was a more eclectic mix of approaches, and a general denial of the completeness of classic Mendelism (nuclear dominance). In the Soviet Union during the Lysenko era, a particular non-Mendelian approach dominated. I will make use of some excellent recent historical work on Mendelism (e.g., Robert Kohler 1994) and alternatives to Mendelism (e.g., Richard Burian et al 1988 on French genetics, Jonathan Harwood 1993 on German genetics) as well as general overviews by Peter Bowler (1989) and Jan Sapp (1987).

As with late nineteenth century evolutionary biology, early twentieth century genetics was pursued with a wide range of decision vectors. The normative upshot, however, was different, as an improper linear analysis of decision vectors will show.

The decision vectors are as follows:

a. Mendelian genetics became successful in the USA at least in part because Thomas Morgan, who developed it, used his position and excellent political skills, as well as the support of his department and university (Columbia and later Cal Tech) to create a new discipline, with its own journal (*Genetics,* starting in 1916) publishing only work in classical genetics. (Non-empirical decision vector, Mendelism +)

b. The university system in the USA was expanding, and open to the creation of new disciplines. Genetics, and Mendelism in particular, was easily accepted as a new discipline. (Non-empirical decision vector, Mendelism +)

c. In Europe, the university system was more rigidly structured, and founding a new research field was almost impossible. Genetic research of all kinds suffered. Mendelism suffered disproportionately, because it is more discontinuous with past traditions.[13] (Non-empirical decision vector, Mendelism −)

d. The eugenics movement, especially in Britain and the United States during the early part of the century, favored Mendelism because of its assumption of "hard heredity" (heredity controlled by deterministic factors), success in breeding experiments and denial of the inheritance of acquired characteristics.[14] The eugenics movement was ideological, claiming that the existing human social order is "natural." Many Mendelians were also eugenicists (e.g., Davenport, East, Castle, Muller, Little, Pearson, Haldane, Julian Huxley) and some prominent Mendelians, although not eugenicists, were sympathetic to or tolerant of it (e.g., Bateson, Morgan). Castle's text *Genetics and Eugenics* was the most widely used college text on genetics from 1916 until the early 1930s, going through five editions. (See Kevles, 1985) (Non-empirical decision vector; Mendelism +)

e. The Morgan school and others producing results in classical genetics in the USA were well funded by plant and animal breeders. This was because the traits investigated by classical genetics were of commercial use. This funding increased scientific work on Mendelian inheritance. (See Rosenberg 1976) (Non-empirical decision vector; Mendelism +)

f. Choice of experimental organism also contributed to the dominance of nuclear models of inheritance. As Kohler (1994) has shown, *Drosophila* was bred for traits that conformed to Mendelian theory, and was relatively easy, inexpensive and fast to work with. Moreover, the early Mendelians (especially those in the fly labs at Columbia and Cal Tech) were generous about donating samples of their carefully bred stocks to other researchers who could then set up genetics research inexpensively. *Drosophila* results were thus cognitively more salient and available. (Empirical decision vector; Mendelism +)

g. As Keller (1985), Harwood (1993) and Gilbert et al (1988) have claimed, the hierarchical organization of the cell presupposed by classical Mendelian genetics (and also later by the "master molecule" theory of DNA) was appealing for ideological reasons. Hierarchical organization mirrors the structure of the traditional family, of traditional relations between men and women, and of traditional societal and organizational structures. It was uncommon for a geneticist to be influenced by more egalitarian ideologies, although this did occasionally happen and is documented in Gilbert et al (1988) and Harwood (1993). In such cases, egalitarian ideologies were the background for theories about cellular organization in which the cytoplasm has a more active role. (Non-empirical decision vector; Mendelism +)

h. Mendelism, early on, was incompatible with most evolutionary biology. Mendelian variations were thought to be so small that they rarely conferred adaptive advantage, and even when they did, it was thought that they would be "washed out" in later breeding. There was no known mechanism for development of new traits, since Mendelism was a theory of "hard heredity." Evolutionary biologists thus tended not to be Mendelians. (Empirical decision vector; non-nuclear +).

i. Later on, however, evolutionary biologists were won over by work on mutations and population genetics (these were later developments in Mendelism) by Fisher, Haldane, Dobzhansky, Julian Huxley, and Sewell Wright. Even though some of the problems regarding the adaptive advantages of small variations had not been solved, the problems of novelty and "washing out" were resolved with the discovery of new mutations and the development of population genetics. Most evolutionists joined the synthesis of natural selection with Mendelism by 1940 and became the staunchest defendants of Mendelism. (Empirical decision vector; Mendelism +)

j. Mendelism separated the previously joined fields of heredity and development, treating the former independently of the latter. Classic Mendelian theory is a somewhat abstract theory about the relation between unobserved factors (genes) and adult phenotypic characteristics. The theory had much empirical success, and therefore was appealing to those of an empiricist or pragmatic bent (which, therefore, is another empirical decision vector). Those who had theoretical qualms, and wanted a theory with more completeness and breadth, were dissatisfied with Mendelism. Harwood (1993) has shown that in the USA and in the UK, researchers were more likely to be empiricist and pragmatic and favor Mendelism, while in Europe it was the reverse. Intellectual styles affected the reaction to Mendelism, as they have to many other issues. (Empirical decision vector, Mendelism +; Non-empirical decision vector, non-nuclear +)

k. Embryologists, who are concerned with growth and cellular differentiation, were especially dissatisfied with Mendelism. They dwelt on the paradox of Mendelism that different cells develop from the same genetic material. (Richard Burian (1986) has called this "Lillie's Paradox" from Frank Lillie 1927, p. 367). They were well represented among researchers on cytoplasmic inheritance (e.g., Lillie, Loeb, E. B. Wilson, Harrison, Conklin).[15] Cytoplasmic inheritance explains growth and differentiation because the cytoplasm is itself anisotropic. Moreover, different kinds of cells have different kinds of cytoplasm. Differentiation thus can come from prior differentiation. This explanatory power is an empirical success of theories of non-nuclear inheritance. (Empirical decision vector, non-nuclear inheritance +)

l. In France, research on non-nuclear inheritance was encouraged because of neo-Lamarckian chauvinism. It was though that nuclear inheritance was "hard" and non-nuclear inheritance more susceptible to environmental influences. Later on, after World War 2, cytoplasmic inheritance was investigated as a competitive strategy with the USA and the UK (see Sapp 1987, chapter 5). (Non-empirical decision vector, non-nuclear inheritance +)

m. Work on the inheritance of acquired characteristics through cytoplasmic mechanisms was favored in the Soviet Union during the Lysenko era (1940s to 1960s). Again, this was because it was thought that non-nuclear inheritance allows for environmental effects on heredity. Marxists, on political grounds, preferred such neo-Lamarckian ideas. Mendelian genetics was denounced as a fascist view, because it postulates "hard" heredity. (Non-empirical decision vector; non-nuclear inheritance +)

n. The work in the USA of Sonneborn, Jollos and Goldschmitt on cytoplasmic inheritance and cytoplasmic role received less attention and respect than it deserved, in part because they were Jewish, and two of them (Jollos and Goldschmitt) recent immigrants from Europe. Sonneborn failed to secure a position at Johns Hopkins because of anti-semitism, and went to Indiana University, which at that time had much less prestige (see Sapp 1987, pp. 111–112). Jollos never re-established his work in the USA, suffering from personality conflicts and anti-semitism. (Non-empirical decision vector, non-nuclear inheritance −)

o. The Cold War, and Lysenko's commitment to cytoplasmic and Lamarckian models of heredity, added to the rejection of non-Mendelian genetics in the USA (Sapp 1987, chapter 6). Sonneborn, and others working on non-nuclear inheritance, came under suspicion of being Communists, and they ended up emphasizing the mainstream aspects of their work. (Non-empirical decision vector, non-nuclear inheritance −)

Summing the decision vectors:

Empirical Decision Vectors

Mendelism	+3
Non-nuclear inheritance	+2

Non-empirical Decision Vectors

Mendelism	+5
Non-nuclear inheritance	+2

The empirical decision vectors are roughly equally distributed. Empirical successes were on both sides. Mendelian inheritance was found in about a thousand cases, whereas non-nuclear inheritance was found in twenty or thirty cases. Non-Mendelians thought that the reason for this is that Mendelian inheritance governs trivial traits such as height and eye color whereas non-Mendelian inheritance is responsible for basic metabolic and developmental processes for which variation usually proves fatal. They had experimental support for this idea: many cases of cytoplasmic inheritance involve photosynthesis, respiration and reproduction. Comparing empirical success of the two approaches in terms of the *number* of demonstrated inherited traits is thus inappropriate. In fact, I see no useful way to rank the empirical success of the two approaches. So equal distribution of empirical decision vectors seems equitable enough.

Non-empirical decision vectors were unequally distributed, with significantly more going to Mendelism. This suggests that Mendelism got more than its fair share of research effort.[16]

4 The Continental Drift Dispute, 1920–1950

A major historian of the history of continental drift, Henry Frankel, claims that "continental drift was not . . . a scientific controversy having significant political, economic or social aspects" (1987, p. 203). Frankel is not alone in this assessment; others who have analyzed the history of continental drift, e.g., Giere (1988), Laudan and Laudan (1989) and Oreskes (1999) have come to similar conclusions. Even those who actively went looking for "social" factors, e.g., Le Grand (1988) and Stewart (1990) found few "external to science" factors such as extra-scientific politics, ideology, etc. Certainly, the case of continental drift differs from the previous two cases, nineteenth century evolutionary biology and early twentieth century genetics in having fewer such "external" factors.[17] This case study and the next (cancer virus research) are chosen for their relative lack of "external to science" factors.

It is just as important, however, to explore the decision vectors affecting these cases. They are as plentiful, and as in need of analysis, as for the previous two cases. Fewer ideological and political factors do not imply anything about the number of empirical or non-empirical decision vectors in the case or the "scientific rationality" of the case.

After the publication of Wegener's *The Origin of Continents and Oceans* (first edition in German, 1915, revised and translated into most languages by the early 1920s), there were three theories available to geologists: permanentism, contractionism, and drift. Permanentists claimed that the Earth's crust has only changed gradually, and without significant continental or oceanic change. It followed from the principle of isostasy (then widely accepted) that continental masses could not sink, nor ocean floors rise. A few land bridges of narrow dimensions were proposed to account for paleontological similarities. Contractionists believed that the earth is cooling, and that this can lead to significant continental change, especially to the wrinkling of the earth's surface and thus to mountain formation (orogeny) and some changes in the proportion of ocean to dry land. Finally, those who accepted Wegener's hypothesis of continental drift believed that the continents had broken up from one, or perhaps two, original large landmasses at the South Pole (or both poles) and drifted towards the equator, leaving the proportion of ocean to dry land the same, and explaining the geological and paleontological similarities between formerly adjacent landmasses.[18]

Each of these theories had supportive data, and each had contrary data as well as significant physical difficulties (as discussed in chapter 2). Nevertheless, many geologists took sides in the debate, and decision vectors influenced their decisions. (Also, many geologists had no opinion one way or another: the debate did not impinge on their work, and there were no strong external reasons for taking sides.) As with the previous cases, I list the decision vectors, in no particular order, classify them as empirical or non-empirical, and record with a + or − sign the theories whose evaluation they affected.

a. In the United Kingdom, Australia and North America, geologists were reluctant to engage in too much theorizing or speculation, preferring to do field studies in which they stayed close to the data. They called this "the inductive method." Permanentism best suited their methodological preferences because it does not postulate large, and now unobservable, changes during the earth's history. (Non-empirical decision vector; Permanentism +)

b. On the other hand, European geologists had a tradition of theorizing and speculation. They were more likely to consider both contractionism and drift seriously. (Non-empirical decision vector; Contractionism + and Drift +)

c. Lyell's influence was especially strong in the United Kingdom and North America. His "principle of uniformity," Originally postulated in order to exclude geophysical theories inspired by biblical narratives (the so-called catastrophic theories) was often thought to exclude contractionism and drift, also. (Non-empirical decision vector; Permanentism +)

d. Wegener hypothesized that the mechanism for drift of continents is similar to the mechanism for drift of large polar icebergs (Wegener [1915]

1966, p. 37). Wegener had traveled extensively around Greenland, and the motion of icebergs was especially salient and available to him. Reasoning by analogy (which results from the representativeness heuristic) produces the hypothesis of continental drift. However, the physical properties of icebergs and continents are not sufficiently similar that the empirical success of the hypothesis of drift follows from the empirical success of the understanding of icebergs. Yet the analogy with icebergs was powerful for Wegener, at least. (Non-empirical decision vector; drift +)

e. Permanentism and contractionism already had followings; belief perseverance phenomena were in their favor. Belief perseverance phenomena include both empirical and non-empirical decision vectors (prior belief makes confirmatory data more salient and available, and there is emotional resistance to change). The non-empirical decision vectors are discussed in the next section, since they are a product of personality. The empirical decision vectors include the salient evidence for permanentism: geological feature showed only gradual and non-dramatic changes over time (Tuzo Wilson's early permanentism, mentioned in the previous chapter, is an example of this). There was also some salient evidence for landbridges, required to explain paleontological similarities, with Panama and the Bering Strait. The folded structure of mountain ranges was salient evidence for contractionism (but note (g) below: horizontal crustal movements provided a better explanation). (Empirical decision vectors: Permanentism ++ and Contractionism +)

f. Sulloway (1996) has found that birth order (as a proxy for the personality trait of conservatism/radicalism or rigidity/flexibility) is highly correlated with theory choice throughout the debate over continental drift: younger siblings were twice as likely as firstborns to adopt drift. The firstborns were more likely to adopt the established theories of permanentism or contractionism, the laterborns likelier than the firstborns (although not likely overall before 1950) to adopt drift. (Non-empirical decision vectors; permanentism +, contractionism +, drift +)

g. Those who worked on orogeny—principally, those who worked on either the Alps (e.g., Argand), the Himalayas (also, e.g., Argand) or the Caledonians (e.g., Bailey)—were more likely to adopt drift, because the folded structure of these mountains suggested that they were formed by horizontal compression of the continental crust. Carozzi (1985, p. 130) writes, "The concept of drifting continents was such an obvious solution to the spectacular structural display offered by the Alps that it was adopted with essentially no opposition." This was despite the fact that anti-German feelings were so high in Switzerland at the time that "it was strictly forbidden to read in public or in private any material printed in Germany" (Carozzi 1985, p. 129). Even some French geologists who worked on Alpine geology adopted drift, and this was right after World War I. As mentioned in (e) above, contractionism was originally the favored theory by those who worked on orogeny (e.g., Suess) but the forces proposed by contractionists—radial forces—do not account as well for the folded structure of the Alps. In fact, geologists inferred global horizontal forces, sufficient to produce continental drift, from the more local horizontal forces, which account for orogeny. Visual salience also played a role here: those who were directly observing the Alps, rather than just reading about the structures, were more likely to adopt drift. (Empirical decision vector; drift +)

h. Anti-German feeling was strong in many parts of Europe after the end of World War I. In Belgium, France and Italy, especially, there was hostility to drift because Wegener, the most important proponent of drift, was German. This was compounded by the loss of many younger, more open-minded,[19] geologists in the war. Anti-German feelings also ran high in Switzerland, but there, concern with orogeny outweighed national sentiments. External political factors were thus not completely absent from the continental drift debate (Non-empirical decision vector; drift −)

i. Those geologists who had traveled and made observations in the southern hemisphere were far more likely to choose drift than those who had not. Geophysical and paleontological data in the southern hemisphere are especially well explained by the hypothesis of drift. (Only much later were northern hemisphere data supportive of drift produced.) Those for whom the southern hemisphere data was available and salient were more likely to choose drift. Many Dutch geologists (e.g., Molengraaff, Brouwer, van der Gracht) who had field experience in Indonesia and South Africa quickly adopted drift. Holmes, who was one of the few

British geologists to take drift seriously, had field experience in Mozambique and Burma. In South Africa, Du Toit, and later his student King, embraced the drift hypothesis and contributed new evidence in support of it. Du Toit's observations were in both South Africa and Australia. Daly (the Harvard geologist who was one of the few in North America to accept drift) regarded Du Toit as "the world's greatest field geologist" (quoted in Le Grand 1988, p. 82). Daly also had southern hemisphere experience (he worked in South Africa with du Toit, although before du Toit accepted drift). Some Spanish and Brazilian geologists were also sympathetic to drift, although this was far from unanimous; probably salient data from just one continent (South America) did not suffice. Those Australian geologists[20] (e.g., Wade) who were willing to speculate were often sympathetic to drift, especially if they worked around the "Wallace Line": the point in Indonesia which, seemingly arbitrarily, is the dividing line between Australian and Asian features. (The explanation of the Wallace Line is that Australia has moved up to Asia only recently.) Giere (1988, p. 240) mentions two quantitative studies (by Schlosser and Stewart) in which southern hemisphere field experience approximately doubled the likelihood of a geologist accepting drift. (Empirical decision vector; drift +)

j. Arthur Holmes, the leading authority on radioactive dating from about 1913, was led by considerations of radioactive heat to propose that there are convection currents in the mantle, with drift as a possible consequence. This theory was, in fact, an anticipation of Hess's later theory of seafloor spreading. It involved making assumptions about the distribution of radioactive material. These assumptions were somewhat arbitrary, as even Holmes acknowledged (Le Grand 1988, p. 116). Holmes presented his ideas about drift in his text *The Principles of Physical Geology,* which was used by generations of British students, teaching them about drift and predisposing them to take drift seriously. (Non-empirical decision vector; drift +)

k. Harold Jeffreys, the British geophysicist, was also led by considerations of radioactive heat. He assumed, however, that radioactive material was confined to the crust of the earth, so that the earth as a whole was cooling and contracting. Jeffreys was strongly opposed to drift. He used his formidable mathematical skills to argue this, and impressed many. Mathematical techniques have high prestige in the scientific community generally, and geology is no exception. Of course, his arguments were no better than the assumptions fed into them. (Non-empirical decision vector; contractionism +)

l. At the time of Wegener's work, longitude measurements appeared to suggest that Greenland was drifting eastward, at a speed of about 11 meters per year. These measurements were always thought to be inexact, but Wegener still found them important and cited them as supportive of drift. Moreover, as Carozzi (1985) has noted, Scandinavians were favorable to drift because of these geodetic measurements. Presumably, the evidence was especially salient to them because of its proximity. By the time this evidence was discredited, it had done its work of persuasion. (Non-empirical decision vector; drift +)

m. In every country, there were regional influences of close peer pressure and local authorities. This increased regional differences, so that each theory gained additional adherents in particular settings. (Non-empirical decision vectors; drift +, contractionism +, permanentism +)

An improper linear analysis of empirical and non-empirical decision vectors, taken from the above account:

Empirical Decision Vectors:

Drift +2

Permanentism +2

Contractionism +1

Non-empirical Decision Vectors:

Drift +5

Permanentism +4

Contractionism +4

This is a fairly equitable distribution of empirical decision vectors, and equal distribution of non-empirical decision vectors, indicating appropriate distribution of research effort. (Note that a whiggish analysis, bifurcating options into "drift" and "anti-drift" would suggest that drift was given inadequate attention.)

5 Cancer Virus Research

Relative absence of ideological and political factors does not generally produce appropriate distribution of research effort. At any rate, there is no reason to think that it is *likelier* to produce appropriate distribution of research effort than cases where ideological and political factors play a major role. Here is a case which, like the case of continental drift, is relatively free of ideological and political factors,[21] but, unlike that case, had a poor distribution of research effort.

I draw from Daniel Kevles's account in which he concludes:

> Unlike earlier episodes in the history of science, the resistance [to cancer virus research] originated in neither religious nor ideological prejudice. It derived from the skepticism of a professional community of biomedical scientists whose beliefs were grounded in available laboratory evidence. (Kevles 1995, p. 106)

Although Kevles does not make particular normative claims, the tenor of this remark suggests that he thought that the skepticism was appropriate and that therefore the resistance to work on cancer viruses was also appropriate. Although I draw on Kevles's historical account, I will draw different normative conclusions.

Peyton Rous discovered cancer viruses in 1909: he discovered that sarcomas in chickens could be transmitted by a non-filterable (and therefore non-bacterial and non-cellular) agent. Rous received the Nobel Prize for his work much later, in 1966. Other researchers showed a few cases of viral transmission of cancer in different animals. Johannes Fibiger had some promising results with the transmission of stomach cancer in rats before World War I, and Richard Shope in the 1930s showed the viral transmission of malignant papillomas in rabbits. But these results were regarded as anomalies, and as having nothing to do with cancer in humans. Kevles writes, "By the 1930s, the theory that at least some cancers were caused by viruses had fallen into deep disrepute . . . , and those who held to the theory risked their scientific reputations" (1995, p. 76).

In 1929, the Jackson Laboratory was founded by Clarence Little, who was the managing director of the American Society for the Control of Cancer (predecessor to the American Cancer Society). Little began a campaign for public education about the disease. He emphasized the role of heredity in cancer. This was in part because he was a eugenicist, and in part because he did not want to create fears that cancer is contagious. Strains of mice were bred that had a high incidence of cancer. In 1936, however, a Jackson scientist, John Bittner, discovered that there is a cancer causing agent in mice milk. He was hesitant to challenge the genetic model, and simply called the agent a "milk factor." In 1937, Little helped create the National Cancer Institute, which quickly concluded that viruses and other microorganisms could be disregarded as causes of cancer. Work on the "milk factor" continued, but it was not until the late 1940s that a viral interpretation achieved general respectability.

Ludwik Gross, a Jewish refugee who arrived in 1940, experimented with the transmission of leukemia in mice (he obtained the strains from John Bittner). He published his positive results in 1951 and 1952, but most cancer researchers did not take him seriously. They had difficulty reproducing his results. Although no one knew it at the time, Gross obtained his results because he injected the virus into a strain of mice that were especially vulnerable to it. In the mid-fifties, Jacob Furth repeated the experiments with the same strains of mice, and replicated Gross's results. Because Furth was well respected, his results were believed. By the early 1960s, research on animal tumor viruses was flourishing again.

At this point, molecular genetics had become an established field. Viruses were demonstrated to be composed of either DNA or RNA and a protein coat, and to replicate by co-opting the DNA of the host. Understanding the mechanisms of viral replication was, reasonably, thought to lead to a greater understanding of cellular processes. The RNA viruses, in particular, presented a challenge to the "central dogma" that DNA produces RNA but not vice versa: for RNA viruses to be incorporated into the cellular DNA, the reverse process had to occur. Howard Temin suggested this, and was widely dismissed. Eventually however, in 1969, he (and at the same time, independently, David Baltimore) discovered that the reverse process occurred under the action of an enzyme, "reverse transcriptase."

Still, this research was thought irrelevant to human cancer. Only the Epstein-Barr virus—a DNA virus—had been shown to cause cancer (Burkitt's lymphoma). RNA viruses (retroviruses) were not thought even to infect humans. By the early eighties, however, some human retroviruses were shown to cause cancer (leukemia), and before long, the retrovirus causing AIDS was discovered. Although viruses do not cause most human cancers, the research on cancer viruses led to a greater understanding of malignancy, because it turns out that the mechanism by which cancer viruses cause malignancy (formation of an oncogene) is an example illustrating the process of oncogenesis, which is a general stage in the development of cancer. Michael Bishop's and Howard Varmus's work in the 1970s was central in the discovery of oncogenes.

In this debate, two theories about the cause of cancer were dominant: the theory that cancer is cause by a virus, and the theory that cancer is "genetic." The decision vectors are as follows:

a. The few findings of viruses that caused cancer were dismissed, in part because there were few additional findings, but also because Mendelism in the USA was dominant, and created a presumption in favor of genetic models. The eugenics movement gave an additional push in the same direction. (Non-empirical decision vector; genetic model +)
b. In the new climate of instituting public education and lack of hysteria about the disease, in which Little was so prominent, the viral model of cancer was discredited so that there would be no fear of contagion. (Non-empirical decision vector; viral model −)
c. Gross's work was dismissed in part because, as an immigrant, he was marginal in the scientific community. (Non-empirical decision vector; viral model −)
d. Furth's authority promoted the acceptance of Gross's work (Non-empirical decision vector; viral model +)
e. Ideas about retroviruses were dismissed because they contradicted the then popular "central dogma." This is a non-empirical decision vector because of the role of non-empirical decision vectors in the popularity of the "central dogma." (Non-empirical decision vector; viral model −)

Empirical decisions vectors have not been detailed here. They are undoubtedly present (at least in the form of salience and availability of particular evidence to particular investigators). There is every reason to think that they were equitably distributed. Little history has been done on this case, however, and empirical decision vectors have not been suggested or discussed by others. Certainly, both theories had empirical success.

An improper linear model of the non-empirical decision vectors in cancer virus research yields the following result:

Non-empirical Decision Vectors

Viral Model	−2
Genetic Model	+1

This indicates an inequitable distribution of non-empirical decision vectors.[22]

6 The "Invisible Hand of Reason"

This chapter has shown that there is no "invisible hand of reason" that guarantees a good distribution of research effort. Cases in which there are few "external" decision vectors are no different, in this respect, from cases in which there are many "external" decision vectors. This implies that there is a general need for a normative epistemology of science that applies at the social level. A descriptive epistemology is not enough. I have begun this normative epistemology here, suggesting that empirical decision vectors be equitably distributed and non-empirical decision vectors be equally distributed. Implicit is the further requirement that theories under consideration have some robust and significant empirical success. Since empirical success is a primary goal of science, theories without such success do not have scientific merit.[23]

6

Consensus

Not the violent conflict between parts of the truth, but the quiet suppression of half of it, is the formidable evil.

John Stuart Mill, *On Liberty*

1 Background

Mill thought that consensus, in science as well as other forms of inquiry, is an obstacle to progress, rationality and truth.[1] He had three general epistemic reasons for resisting consensus. First, since humans are fallible, a dissenting opinion might be the correct one. Second, opposed views might each be partly true and partly false, so that consensus on one view would result in the loss of truth. And third, Mill claimed that even where there is consensus on a true theory, and no truth is lost by the rejection of competing theories, there is epistemic loss of "the clearer perception and livelier impression of truth, produced by its collision with error" (Mill 1859, p. 143). A better understanding of the meaning of a true theory is reached if one knows the pro and con arguments (Mill 1859, p. 170).

Mill's views about dissent and consensus have never achieved much popularity in philosophy of science. Realists and antirealists alike view consensus on a particular theory, method or practice as the rational endpoint of episodes of scientific change. Kitcher's 1993 statement is typical and contemporary:

> When disputes are resolved, when all the variants but one in some part of some component of individual practice are effectively eliminated, there is a change in *consensus practice*. If we are to understand the progress of science, we need to be able to articulate the relations among successive consensus practices." (Kitcher 1993, p. 87)

Dissent is seen as a necessary period when a thousand theories can bloom (see, e.g., Hull 1988, p. 521), yet as a *temporary* stage on the way to consensus. Realists associate consensus with truth; antirealists with greater success or rationality.

The situation is similar in the wider science studies field, and in science policy. Consensus is viewed as an important (if not rational) endpoint of inquiry. Collins and Pinch express a typical view when they write, "Science works by producing agreement among experts" (Collins and Pinch 1993, p. 148). The recent proliferation of medical consensus conferences, in this country under the auspices of NIH and AHCPR, suggests the importance given to consensus formation in medicine.

Paul Feyerabend is, of course, a major recent exception to the consensus on consensus. As Lisa Lloyd has argued (1997), Feyerabend saw his own work as an exemplar of Mill's epistemic proposals. Interestingly, for all the influence of Feyerabend on contemporary history and sociology of science, the production of narratives ending in consensus continues in those fields as well as in philosophy of science. This is especially curious because "strong programme" sociology of science is skeptical of claims to both rationality and truth—and claims to rationality or truth are the typical links to an expectation of consensus. Perhaps sociologists and historians have been influenced by philosophers to focus on consensus. Perhaps, also, a social enterprise in which there is widespread agreement is easier and more familiar to conceptualize and explain. Indeed, the same assumption of consensus—univocality—is traditional in cultural anthropology. In addition, I suspect that narrative forms in which stories have an end (consensus is easily seen as an endpoint) and in which there are winners and losers (the theory on which there is consensus is often seen as the winner) are being assumed here.

Another notable recent exception to the consensus on consensus is the work of Helen Longino (1990), who argues for scientific pluralism as a goal in some fields of science, not just as a means to finding "the best theory" but also as an appropriate result of inquiry. Her work has had some impact in philosophy, e.g., on the recent ideas of John Dupre and Ronald Giere, but the impact is not widespread.

Consensus is even more important to the agenda of most philosophers of science than it was in the past, before the Kuhnian revolution. Kuhn convinced many philosophers of science that there are episodes of serious dissent in science, and that ideological and psychological differences among scientists cause them to take different sides in a debate. Historically sensitive philosophers of science (for example, Giere, Hull, Kitcher, Thagard) found a way to absorb and counter this apparent challenge to scientific rationality. They view the period of dissent as a period of investigation of alternatives, and ideological and psychological factors as ways of distributing cognitive labor among the alternatives. Then, consensus takes place on the most successful theory, which scientists choose on rational grounds such as "most empirical success," "most explanatory coherence," etc. Frequently, there are also claims about the association of consensus with truth (or, partial truth, resemblance of the theory to the world, etc.).

Similarity to the traditional context of discovery/context of justification distinction is strong.[2] "Context of discovery" was thought of as the opportunity for individual creativity, in which a variety of psychological processes, including dream associations, might play a role, and assessments of rationality are irrelevant. "Context of justification" was thought of as the time when the theories developed during the time of discovery are rationally assessed. Only when a theory survives justification was it thought to be an acceptable scientific theory; until then, it was thought to be indistinguishable from flights of fancy, e.g., literature. On this understanding, dissent is the contemporary, post-Kuhnian, "context of discovery" where "anything goes" and consensus is the contemporary "context of justification." Thus what is taken to be the context of discovery has expanded, to include more of the early stages of development of a theory, including its initial assessment and reception in the scientific community.[3] Often (e.g., in the writing of Giere, Hull, Kitcher) this is termed an "evolutionary" account of scientific change, where dissent is seen as due to "random variation" and consensus due to "survival of the fittest" theory.[4]

For all, dissent is seen as the stage of competition, consensus the stage when there is a winner of the competition. Dissent that does not resolve itself for a while is frequently regarded as incomplete or immature science. Kuhn's category of pre-paradigm science is an early theorization of this view. Dissent is seen as a stage in need of, or inevitably headed towards, resolution in consensus. Our cultural obsession with competition, winners and losers (e.g., military and sports discourse) provides an ideological background that resonates with these narratives of consensus in science studies and, most likely, contributes to them.[5]

There is another dimension to the consensus on consensus. Accounts of consensus, whether philosophical, historical or sociological, tend to explain consensus as the product of a single central or universal cause. Thus, not only is there a consensus on the normativity of consensus, there is consensus on a type of explanation for it. Philosophical accounts often attribute consensus to a single universally shared reason or justification, such as that one theory has the most predictive power, the most problem solving ability, the most psychological coherence, or the most unifying explanatory power. Sociological accounts frequently point to central social processes, such as the triumph of the bourgeoisie, using them to explain, for example, the outcomes of the Scientific Revolution and the Chemical Revolution (see, e.g., Shapin and Schaffer 1985 and McEvoy 1983). Explanatory accounts of consensus typically differ from explanatory accounts of dissent, since the latter appeal to a variety of causes (decision vectors) to explain decisions.

Some sociologists (notably Latour 1987) have developed a more sophisticated and decentralized view of social processes, in particular of the ways in which power is exercised. These views are currently too amorphous to be anything other than suggestive, and they lack (deliberately) any analysis that might yield normative assessments. A few philosophers (notably Laudan and Laudan 1989, Husain Sarkar 1983) have also given a distributed account, suggesting that consensus typically occurs when scientists have different reasons for preferring the same theory. This would yield an interesting normative account if it had any descriptive accuracy; as I shall show, the case studies show that consensus (even normatively appropriate consensus) forms in a different way from that suggested by the Laudans and Sarkar.

2 Initial Reflections

Chapters 2 and 3 argued that scientific progress without consensus, as well as without truth, is a common occurrence. The most universal goal of scientific activity is empirical success, rather than truth, and even when truth is a reasonable goal, empirical success is the accompanying goal.[6] It follows that *neither* dissent nor consensus is invariably or intrinsically valuable for scientists; sometimes dissent will maximize empirical success and truth, and sometimes consensus will. In saying this I am not only disagreeing with those who claim that consensus is intrinsically normative, but also with Mill, Feyerabend, Longino, etc., who see intrinsic merits in dissent. Obviously, dissent will be appropriate when different theories have different empirical successes. On the other hand, consensus will be appropriate when one theory has all the empirical success. In the case studies that follow, I shall show that this insight from chapters 2 and 3 is usefully applicable to case studies.

Decentralized and complex accounts of causal processes are becoming popular now in all areas of research, in part because of advances in mathematics and computer science. It is true that centralized (linear, general, etc.) accounts are still easier for us to understand, conceptually and mathematically. This explains, at least in part, their continued dominance. The case studies that follow, however, will show that consensus in science typically occurs in a distributed manner. They also show that as many decision vectors are involved in coming to consensus as are involved in producing dissent. Any normative account needs to accommodate these findings.

Just two cases studies of consensus are chosen for this chapter: consensus in the 1960s on plate tectonics, and consensus in the 1950s on the "central dogma" that nuclear DNA controls both inheritance and cellular processes. The case studies show two things. First, consensus on plate tectonics was especially normatively appropriate, while consensus on the "central dogma" was less so. Second, in both cases, consensus was brought about by a variety of distributed processes in which decision vectors are as varied, and as powerful, as they are during times of dissent. Chapter 7 will use these observations to suggest a normative account of consensus, and show how the account—*social empiricism*—usefully applies to a wide variety of cases of both consensus and dissent. According to social empiricism, consensus is no more and no less a "context of justification" than is dissent.

3 Consensus on Plate Tectonics

Between 1958 and 1970, almost complete consensus formed on the mobilist ideas of seafloor spreading and plate tectonics. Recall from chapter 5 that after about 1930 geologists had settled into a protracted pattern of dissent about continental drift, with, for example, southern hemisphere geologists more likely to be mobilists than those who did northern hemisphere fieldwork.

During the 1950s and 1960s, geologists, and especially geophysicists, produced an abundance of new observations and measurements, especially in the areas of oceanography and paleomagnetism. Since this data proved especially relevant to the new dynamics of continental drift (theories of seafloor spreading followed by theories of plate tectonics), many have thought that scientists became mobilists because they were overwhelmed by the range and amount of positive evidence for continental drift. Glen, for example, writes that the "confluence of evidence" from paleomagnetism and oceanography quickly caused consensus on plate tectonics (1982, p. 10). Giere writes that the data in support of drift was so strong that it overwhelmed "nonepistemic values" in individual scientists (1988, p. 227). And Thagard claims that drift was chosen because by the mid-1960s geologists each, individually, saw that it had greater "explanatory coherence" (this includes coherence of observational and theoretical claims) than permanentism (1992, p. 181).

These accounts are historically incorrect. Scientists achieved consensus on mobilism in a more piecemeal and distributed fashion. Menard, one of the geologists involved in the plate tectonics revolution, writes, "It will be a characteristic of this revolution that most specialists were only convinced by observations related to their specialties" (Menard 1986, p. 238). Other historians, for example Le Grand (1988, p. 221), have repeated this observation. The order of consensus formation was paleomagnetists, oceanographers, seismologists, stratigraphers and then continental geologists concerned with paleontology and orogeny. Belief change occurred after observations confirming mobilism were produced in each specialty, and after old data was reinterpreted to be consistent with mobilism, as I shall describe below.

Moreover, in addition to this general pattern, there were patterns of belief change and belief perseverance due to prior beliefs, personal observations, peer pressure, the influence of those in authority, salient presentation of data, and so forth. I shall describe some of these below.

All these patterns can be explained in terms of decision vectors. Moreover, the decision vectors are similar in kind, variety and magnitude to those causing dissent in other historical cases. The result here is consensus because, as I shall show, a critical number of the decision vectors cause belief in plate tectonics rather than its alternatives.

The first set of new data to point in the direction of drift was from continental paleomagnetism. Geophysicists such as Runcorn, Blackett and Bullard in the United Kingdom, Cox, Doell and Dalrymple in the United States and Irving, McDougall and Tarling in Australia discovered, from volcanic rock samples taken from all over the world, that the variation in remnant magnetism is best explained by a combination of continental drift and an apparently random pattern of reversal of the earth's magnetic field. Some paleomagnetists (most famously, Runcorn) espoused drift already in the 1950s, when the data was still uncertain. In 1959, Runcorn argued in *Science* (vol. 129, pp. 1002–11) that drift had occurred—whatever the mechanism might be. By 1960, paleomagnetists concurred that the data in favor of drift was robust. If conclusive data is all that is needed to produce scientific consensus, why did the rest of the geological community not take the data more seriously? Even the oceanographer Menard, who did not change his fixist views until the mid-1960s, writes:

> The classical continental data plus paleomagnetism might have proved drift without the marine observations. The continental magnetic observations were generally viewed as inconclusive at best when the marine data settled the question of whether continents drift. Nevertheless, most of the perceived problems in paleomagnetism were already resolved. (1986, p. 300)

The early paleomagnetic work was actually viewed with suspicion by field geologists, whose methods were more "low-tech" and whose fieldwork was considerably less glamorous (volcanoes tend to be in exotic vacation spots). Any "reasons" that field geologists may have had for regarding the data as inconclusive were certainly clouded and magnified by ignorance and interdisciplinary rivalry. (See, for example, Appendix B in Glen

1982, which contains parodies on the new paleomagnetism written by field geologists as late as 1963.) Field geologists called the paleomagnetists "paleomagicians." It should also be noted that many of the paleomagnetists were British (British scientists had always been more sympathetic to drift than those in the United States) or Australian (thus more likely to support drift because of local, southern hemisphere, data). And, finally, paleomagnetic data were especially salient to the paleomagnetists who produced it; they weighted their own data more heavily than did outsiders. They were inclined to find this data more compelling than geophysical reservations about the plausibility of continental drift. Thus, empirical and non-empirical decision vectors explain the pattern of reception of the terrestrial paleomagnetic data. Paleomagnetists were influenced by salience and availability of data; for others, mere knowledge of the data was not sufficient to overcome belief perseverance and greater theoretical reservations.

By 1960, the postwar oceanographic expeditions (notably from Scripps, Woods Hole and Lamont) had made a number of discoveries, including a pattern of seismically active mid-ocean ridges and peripheral trenches, giant fracture zones separating portions of ridge and trench and the relatively young ages of the ocean floors. In 1960, the oceanographer Harry Hess put forward a model of "seafloor spreading" (a phrase coined by Robert Dietz), which argued that the continents drift over the same trajectories described by Wegener, but by riding passively over seafloors. (Hess, incidentally, had worked with the Dutch geologist Vening-Meinesz, who was a drift sympathizer who worked on Southern Hemisphere materials, in the 1930s.) The appropriate metaphor is a feather stuck on the top of slow flowing tar, rather than an iceberg moving through water. The flowing seafloor, powered by mantle convection currents, is produced at ocean ridges and consumed at trenches. Continents, which are lighter, are not consumed at trenches, but are deformed when they collide with one another or come into the proximity of ridges and trenches. Hess himself called the model "geopoetry" (rather than, e.g., a new theory); he was well aware of the opposition to drift in the United States, so he put forward his ideas with modest confidence.

The combination of the results of the paleomagnetic studies and Hess's seafloor spreading model revived interest in drift. Oceanographers looked to their own data for confirmation or disconfirmation. Vine and Matthews (who were predisposed towards drift because of their respective educational and field experiences) took up the hypothesis of seafloor spreading and applied it to new seafloor magnetic data. During the early sixties, magnetic striations were found on the seafloor parallel and close to ridges. Vine and Matthews (and simultaneously a Canadian paleomagnetist, Morley) hypothesized that seafloor spreading together with reversals in the earth's magnetic field produce this pattern. Their paper (published in *Nature* in 1963) received little attention. In the meantime, the Canadian geophysicist J. Tuzo Wilson became convinced of seafloor spreading after finding that the age of oceanic islands was correlated with their distance from ridges.[7]

On the other hand, Gordon MacDonald, a geophysicist and opponent of drift, argued (1964) that the continents extend too deeply to be carried by the seafloors. Opponents of drift took his arguments seriously. Moreover, some who tested the hypothesis of seafloor spreading obtained contrary data. Notably Ewing, an oceanographer who headed the Lamont laboratory, rejected drift after finding that his data on the thickness of sediments did not match the predictions from seafloor spreading.

Both empirical and non-empirical decision vectors thus mediated the new interest in seafloor spreading. Those who were predisposed towards drift (e.g., because of field experience giving empirical decision vectors, educational training producing non-empirical decision vectors) and those who discovered data in support of drift (producing empirical decision vectors) tended to accept seafloor spreading. On the other hand, those who were opposed to drift (perhaps just because of conservatism, a non-empirical decision vector) tended to welcome arguments against drift. Moreover, those who put drift to the test and obtained negative data weighed that data more heavily than the data in support of drift that they were not directly acquainted with (due to salience, an empirical decision vector). Thus belief perseverance as well as salience and availability of data were important, mostly empirical, decision vectors.

During 1965, Vine, Matthews and Tuzo Wilson (who joined forces) continued to produce ideas and evidence in favor of seafloor spreading. Wilson used recent data on seismic activity to suggest that transform faults offset portions of spreading ridges; this was an important step towards plate tectonics. And Vine, Matthews and Wilson took the 1963 Vine-Matthews hypothesis further, arguing that if there is a constant spreading rate, the magnetic anomalies on either side of ridges should be mirror images of one another and the age of rocks taken from the seafloor (below sediments) should be directly proportional to their distances from the ridge. Moreover, the seafloor anomalies should correspond to the magnetic reversal scales found by continental paleomagnetists.

The magnetic profiles for the East Pacific Rise and the Juan de Fuca Ridge (part of the same ridge systems) were bilaterally symmetrical and matched one another fairly well. When Vine was told by the continental paleomagnetist Brent Dalrymple of a new magnetic reversal event (the "Jaramillo" event), the correspondences looked even better. Meanwhile, the "Jaramillo" event had also been noticed at Lamont, where previously the Vine-Matthews hypothesis had been rejected. Opdyke, the lone drifter at Lamont (he was trained by Runcorn in the UK), identified the Jaramillo event from sediment cores, which also show the magnetic striations. Walter Pitman, who was processing the magnetic data from traverses of their research vessel across the East Pacific Rise, noticed their bilateral symmetry (especially that of *Eltanin*-19). Pitman had read the Vine-Matthews paper of 1963 and saw that his data pointed in the same direction. Faced with skepticism from most of his colleagues, he took his data to Opdyke who was (as Menard 1986, p. 264, puts it) "the resident heretic." Together they interpreted the data, showing the correspondences between all the magnetic patterns. Fred Vine's arrival at Lamont in February 1966 was, as Glen puts it "like a starving man let loose in the kitchen of the Cordon Bleu" (1982, p. 336). He also seized the data, and published separately.

The oceanographer Menard writes of the *Eltanin*-19 profile, "That profile alone was capable of converting to sea-floor spreading almost everyone who was involved enough to appreciate the issues" (1986, p. 256). William Glen, a historian of the plate tectonics revolution, shows the importance of the magnetic oceanographic data by calling his book *The Road to Jaramillo*. Yet, not everyone converted to drift at that time.

At first, the senior scientists at Lamont dismissed the results. Heirzler, Pitman's advisor, as well as Worzel, thought that the results were "too perfect" (Le Grand 1988, p. 215); Heirzler took about a month to change his mind (Menard 1986, p. 335). It was almost a year later before Ewing, the head of Lamont, grudgingly came around to accepting drift, calling it "rubbish" as late as November 1966 (Menard 1986, p. 267). Even then, he continued to raise difficulties with the hypothesis. Lamont had a long track record of opposition to drift; no doubt belief perseverance, pride, authority and such mostly non-empirical decision vectors played an important role in slowing the conversion to drift. The significance of crucial data being produced at Lamont itself should also be noted. Such data was especially salient and available at Lamont (and especially so to Pitman), providing empirical decision vectors in the opposite direction at that laboratory.

Several historians—especially those focusing on the marine discoveries (e.g., Menard, Glen, Giere)—think that it was the linear magnetic anomalies that clinched the case for seafloor spreading and drift. As Menard said, "it was the discovery of magnetic symmetry that tipped the balance" (1986, p. 238). Menard went on to say, "The symmetry of the magnetics was capable of convincing almost all geologists, regardless of specialty, if only they would look, and the symmetry was first demonstrated by Fred Vine in this miraculous year of 1965" (1986, p. 238). If we look at the actual pattern of belief change, however, 1965 stands out as critical only for oceanographers, and moreover, only for the subset of them that were placed well relative to the data (i.e., for whom empirical decision vectors played a crucial role).[8] Even Ewing, Heirtzler, Le Pichon and Worzel looked at the data—but they were not placed well enough to evaluate it positively. Paleomagnetists were convinced of drift much earlier (they could have said of the oceanographers 'if only they would look at the continental paleomagnetic data'!), and seismologists, stratigraphers and continental geologists somewhat later.

As I mentioned above, Menard wrote more accurately of the plate tectonics revolution when he said, "It will be a characteristic of this scientific revolution that most specialists were only convinced by observations related to their specialties" (1986, p. 238). And he followed up with "Specialists would work to be convinced by their special knowledge and the converted would develop remarkably fruitful corollaries about continents and ocean basins" (1986, p. 294).

The next field in which evidence for plate tectonics was produced was seismology. Seismologists, in turn, were persuaded by salient and available data. Wilson's description of transform faults was quickly followed by Lynn Sykes's confirmation of them. Morgan, impressed by the location of transform faults (along great circles), had begun to develop the quantitative apparatus of plate tectonics. Oliver, Sykes and Isacks, taking seismological measurements in the southwestern Pacific during 1964–65, were ready in 1966 to interpret their results. They had inside knowledge of seafloor spreading, transform faults and Morgan's plate tectonics. Their data confirmed the existence of trenches, thus supporting plate tectonics and disconfirming the recently developed mobilist alternative of an expanding earth. In 1967, McKenzie and Parker confirmed rigid plate motions through seismology. Dan McKenzie, a recent PhD, was present at the 1966 Goddard conference (at Columbia), where papers by Vine and Sykes convinced him of seafloor spreading and motivated him to work on plate tectonics.

By this point, the three most powerful oceanographic research institutions—Lamont (Columbia), Scripps (UCSD) and Madingley Rise (Cambridge)—had joined forces, sharing data, ideas and results, despite occasional disputes over priority. Special conferences—sometimes attendance by invitation only—were arranged to spread the word about drift. For example, McKenzie was one of 39 invited scientists to the Goddard conference. The first Gondwana conference was held in 1967 in South America. Also, more papers about drift began appearing at regular conferences such as GSA and AGU. Thus, a mixture of empirical and non-empirical decision vectors (authority, peer pressure and salience) operated in favor of drift. Opponents to drift deplored what they saw as a bandwagon effect (Le Grand 1988, p. 251)

At the special conferences, geologists from other fields—stratigraphy and other areas of continental geology—began to come on board, interpreting and reinterpreting data to support seafloor spreading and plate tectonics. At the 1966 Goddard conference, Marshall Kay (Columbia) and John Dewey (Cambridge) argued that drift could explain pattern matching between the Canadian Appalachians and the British Caledonians. Also Patrick Hurley (MIT) used isotopic dating methods to give more data to support pattern matching between Brazil and

Nigeria. The 1967 Gondwana conference included papers by continental geologists.

Geologists did not, in general, become convinced of drift as soon as they were presented with the evidence for it, or even as soon as they were presented with evidence for drift from their own field in geology. Publication of the Vine-Matthews hypothesis, for example, had little effect. Strong decision vectors came from direct personal contacts within informal networks, the prestige of the Lamont, Scripps and Madingley Rise laboratories (from which most of the results were produced), special conferences at which the invited people were selected for their actual or potential scientific prominence, and the salient manner of presentation of the data—in diagrams (sometimes chalk drawn with two hands, demonstrating symmetry) and models (especially Wilson's paper folded model of transform faults). Eventually, the entire geological community (with the exception of a few isolated holdouts) came on board. Apparently conflicting data (such as the apparently advanced age of some oceanic rocks) were reinvestigated and reconciled (dating methods were refined). There was consensus on plate tectonics.

The consensus did not happen after any "crucial experiment." In fact, no single experiment was "crucial" for more than a group of geologists at a time (for whom that data might be especially salient and/or available). The consensus was gradual, and distributed across subfields of geology, countries, and informal networks of communication. It was not significantly distributed across subgroups having different scientific methodologies (as Laudan and Laudan 1989 argue). Consensus was complete when plate tectonics had all the empirical successes (those that permanentism, contractionism and drift had, and more besides)—but not because any individual or individuals saw that it had all the successes. Moreover, empirical success alone was not sufficient to bring about consensus. Decision vectors—both empirical and non-empirical—were crucial. In chapter 7, the comparative role of empirical and non-empirical decision vectors will be analyzed more closely.

4 Consensus on the "Central Dogma"

In the previous two chapters, I described the history of genetics up to the 1950s. To summarize quickly, I found that Mendelism was the dominant theory, not only because of the experimental successes associated with it, but also because of an unequal distribution of non-empirical decision vectors. Alternative approaches—which all gave the cytoplasm a greater role in matters of inheritance—had not only fewer experimental successes, but also fewer non-empirical decision vectors working in their favor.

During the 1950s, classical (Mendelian) genetics was largely replaced by molecular biology, and the dominant theory in molecular biology—in the United States and in Europe—was "the central dogma" (as James Watson termed it) that DNA controls cellular processes, via messenger RNA and protein synthesis.[9] This theory is the biochemical successor to the Mendelian view that there is nuclear (chromosomal) dominance. And, in fact, there was even more support of "the central dogma" than there had been of Mendelism—enough support that historians often talk of consensus on the central dogma.[10] Textbooks were rewritten to teach "the central dogma" as a basic law of biochemistry. Histories, until quite recently, retold the history of genetics as culminating in Watson and Crick's 1953 discovery of the structure of DNA (a prominent example of the traditional history is Olby 1974).

Such histories have tended to obscure an interesting feature of molecular genetics: its experimental techniques, materials and goals are quite different from those of traditional Mendelian genetics. Molecular genetics was developed in the context of microbiology, in mid-century. Microbiology at first was used to investigate both Mendelian and other kinds of inheritance. One advantage of using microbes is their rapid rate of reproduction, which enables experiments to be done in a matter of hours rather than days (for Drosophila), months (for corn) or even years (for many higher organisms). Another advantage is that many microbes can be cultured in controlled mediums, where their uptake of biochemicals can be measured and general biochemical assays can be performed. It was through such experiments that DNA was isolated (by Avery, Luria and others) and the process of protein synthesis was discovered (by Beadle, Ephrussi, Jacob and Monod, Lwoff and others).

Molecular genetics, in its early days, saw itself as the successor to traditional Mendelism. It inherited the theoretical claim of nuclear monopoly, now taken further in the claim to have reduced chromosomal genes to the biochemistry of DNA and protein synthesis. Experimental techniques from traditional genetics became outmoded; biochemical procedures took over. Many molecular geneticists came from backgrounds in physics and chemistry, rather than biology. This situation pushed into the background both traditional experimental techniques from genetics and research on genetic mechanisms that did not fit into "the central dogma." Excellent work in these areas (e.g., by McClintock, Sager, Sonneborn) was dismissed, often without being understood, and new jobs went to biochemical geneticists. Many of those who made important contributions to biochemical genetics came from the tradition of research on non-nuclear inheritance yet had their work packaged in terms of "the central dogma" (e.g., Nanney, Ephrussi, Jacob and Monod). The mechanisms of feedback regulation of nuclear genes were not seen (as they could have been) as a challenge to the dogma of nuclear monopoly; instead they were simplified (through the work of Beadle) and incorporated in a linear framework (one gene, one protein, one feedback mechanism). No serious attempt was made to understand enough about nuclear regulation to explain cellular differentiation and development of multi-celled organisms (since micro-organisms are unicellular, this concern disappeared). Ephrussi and Sonneborn, both highly deserving, were passed over for Nobel prizes.

Molecular genetics was, and still is, exceedingly experimentally successful. Certainly there were many empirical decision vectors working in its favor in the genetics community. There were, however, other sources of support. Not only were most of the non-empirical decision vectors in favor of Mendelism (discussed in the previous chapters) still operating, now in favor of "the central dogma," but, also, many geneticists had additional metaphysical preferences for molecular and "central dogma" approaches. As Keller (1983), Burian (1996) and others have argued, many early molecular geneticists saw in the "central dogma" a simple, universal law of biology that has the additional virtue of reducing biology to chemistry. The hope was (and still is, among many) to "unify the sciences" by finding fundamental laws in each of them, and reducing those laws to regularities from the most fundamental science (physics) and to regularities of the most basic component ontology (atoms and their component particles). This is a metaphysical view that still influences decisions now, in talk of the genome as "the secret of life" and over-emphasis (and funding) on the human genome project, which primarily sequences the genome.

During the consensus on the "central dogma" in the 1950s and 1960s there were several empirical successes challenging the "central dogma" that were overlooked or dismissed. Not only was earlier work on cytoplasmic inheritance downplayed, but also new work on cytoplasmic inheritance, complex nuclear interactions and inheritance of supra-molecular structures was ignored. This new work built on previous work in traditional genetics on cytoplasmic inheritance. Sonneborn and Ephrussi were important and continuing influences.

Ruth Sager and Nicholas Gillham's work on cytoplasmic inheritance in Chlamydomonas yielded around 40 examples of non-chromosomal inheritance by the mid-1960s (Sapp 1987, p. 207). This work was not taken seriously, in part because it did not fit in with the "central dogma," in part because it used traditional genetic techniques rather than the newest biochemical approaches and in part because of sexism. As Sapp reports (1987, p. 206), "there was much gossip about 'Ruth's defense of the egg'." Cytoplasmic inheritance was not accepted until it was shown, a few years later, by new biochemical techniques, that there is DNA in cytoplasmic organelles such as mitochondria and chloroplasts.

Barbara McClintock showed by 1948 that portions of chromosome can move from one location (on one chromosome, or between chromosomes) to another, and that this can affect the expression of genes that do not move. This happens on a regular basis (i.e., it is not a random occurrence). She showed—using traditional genetic methods of breeding, phenotypic observation and observation of stained chromosomes through a light microscope—that this happens during mitosis. Her experimental organism (as almost always) was maize. She presented her work in the early 1950s; the overwhelming majority of geneticists greeted it with incomprehension and, rapidly, dismissal (see Keller 1983). The difficulty was partly the complexity of her data and reasoning; but perhaps most of all the difficulty was that her work did not fit into the simple framework of Mendelian genetics or "the central dogma." Moreover, the techniques of microbiology, molecular genetics and biochemistry rapidly took over in the late 1940s and early 1950s, and few of the new geneticists were able to understand—or find salient—results from maize genetics. It was also easy for geneticists to observe McClintock's intellectual and social isolation and dismiss her, as one prominent geneticist did, as "just an old bag who'd been hanging around Cold Spring Harbor for years" (reported in Keller 1983, p. 141). The sexism that had plagued her career from the beginning now operated not just to deny her professional opportunities (jobs, tenure, etc.) but also to deny her fair reception of her work. It was not until the mid-1970s that "jumping genes" were found in bacteria, through the techniques of molecular biology, and interest developed in McClintock's work. The importance of transposition is still controversial; few share McClintock's view that they are responsible for much of cellular differentiation and play a large evolutionary role. Ultimately (1983) McClintock was awarded the Nobel Prize for her discovery of transposition.

Sonneborn's work on inherited supramolecular structures was almost completely ignored. In a biochemical and reductionist era when cells tended to be viewed as 'bags of chemicals', the importance of the arrangement of these chemicals was not seen. This approach was reinforced by the tendency to work on viruses and bacteria. In viruses, at least, instructions for assembly during replication may not be needed. Sonneborn, working on Paramecium (a single celled ciliated protozoan with much visible internal structure), was keenly aware of complex cytoplasmic structure. He (and also Ephrussi) argued for the importance of cell polarity in cellular replication and differentiation. Sonneborn showed that the cortical structure of Paramecium has organizational characteristics that are transmitted without nuclear involvement. He repeatedly said, "Only a cell can make a cell." Even today, this tends to be overlooked—e.g., in discussions of cloning technology.[11]

The story of consensus on "the central dogma" in the 1950s and 1960s is similar to the story of consensus on plate tectonics, in some respects. Decision vectors, some empirical and some non-empirical, influence all decisions. Consensus occurred by a distributed process in which different individuals joined the consensus in different ways—some by the salience of their own experiments, some by peer pressure, some for metaphysical reasons, etc. Non-empirical decision vectors played a more determining role in the consensus in genetics, however. In addition to the decision vectors in favor of nuclear monopoly discussed in the previous chapter, there was a hope that the science of genetics would be unified with the physical sciences by being grounded in biochemical laws (a non-empirical decision vector: metaphysical preference). Thus, the imbalance of non-empirical decision vectors, present before consensus, is exacerbated through it. This did not happen during consensus on the plate tectonics revolution. Chapter 7 will elaborate on this observation.

Consensus on "the central dogma" had some obviously negative results: the important work of Ephrussi, Sonneborn, Sager, McClintock, etc., was ignored, misunderstood or downplayed. Thus, the consensus was not on a theory with all the available empirical successes; indeed, the consensus got in the way of recognizing and disseminating some empirical successes.

Someone might argue[12] that consensus on "the central dogma" was fine: simple ideas must come before complex ones, and so the consensus was psychologically appropriate as part of the process of discovery. According to this view, textbooks present "the central dogma" because it is pedagogically wise. A variant of this argument is the view that "the central dogma" is a good first approximation to the truth.

I regard this line of argument as *ex post facto* rationalization. Whose psychological abilities are important here? Sonneborn, Ephrussi, Sager, Nanney and McClintock were perfectly capable of creating initial theories that were more complex than "the central dogma." It is not difficult to come up with a counterfactual scenario in which there is *not* consensus on "the central dogma" and progress is *better,* because the important work of Sonneborn, Sager and McClintock is absorbed earlier. If, as I have argued, empirical success is a primary goal of science, consensus on the "central dogma" is an unnecessary compromise.

5 Comments

These examples suggest that decision vectors causing consensus are no different in kind or magnitude from decision vectors producing dissent. They are distributed rather than central or universal, "hot" cognitive as well as "cold" cognitive, social as well as individual, and both empirical and non-empirical. Obviously, in some cases the complex causal dynamics cause consensus, and in other cases they cause dissent; the difference is due to some as yet unanalyzed difference in distribution or interaction of decision vectors.

The genetics case suggests that consensus is not always conducive to empirical success or truth. Empirical successes were ignored or devalued because of the widespread consensus on the "central dogma." And the central dogma is seriously inaccurate.

Consensus in science is typically "shallow" in that scientists might agree on a theory, but for different causes or reasons. A "deeper" consensus would involve the same pathways to a conclusion. Probably, over time, a consensus becomes deeper simply because successive generations of scientists are taught accepted theories in the same ways.

Chapter 7 will take these issues further by giving a normative account of consensus that will elaborate on the important differences between the case of plate tectonics and the case of consensus on the "central dogma." This normative account—an extension of the *social empiricism* that was begun in chapter 5—will then be used more generally, for other cases of consensus.

According to the reflections of this chapter, and chapters 2 and 3, there is nothing intrinsically desirable about consensus. Sometimes consensus is normatively appropriate, and sometimes dissent is normatively appropriate. Consensus is not a *telos* of science that should shape either normative or descriptive accounts. It follows that description and normative assessment of the breakdown of consensus are as much of interest as the processes of coming to consensus. Thus, examples in chapter 7 will also include cases of the breakdown of consensus.

7

Social Empiricism

It's *Not* the Thought that Counts.
Vivien Rothstein

1 Statement of *Social Empiricism*

One of the conclusions of chapter 6 is that there is much similarity between dissent and consensus. The causes of each are the same: decision vectors, both empirical and non-empirical, in aggregate, sometimes produce dissent and sometimes produce consensus. The goals of each are the same: maximization of empirical success and truth. And neither dissent nor consensus is invariably, intrinsically, or even typically normatively appropriate. Cases of dissent and consensus were described in chapters 5 and 6 in which the distribution of research effort was normatively inappropriate, as well as cases in which the distribution of research effort was normatively appropriate. Decision vectors of all kinds are always present, which evokes the "symmetry thesis" of SSK although not the epistemic anarchy that has been presumed to follow from that thesis.

In chapter 5, a normative account of dissent was given. This specifies the conditions under which dissent is normatively appropriate, and also specifies how research effort is best distributed. There were three conditions given:

1. Theories on which there is dissent should each have associated empirical success.

2. Empirical decision vectors should be equitably distributed (in proportion to empirical successes).

3. Non-empirical decision vectors should be equally distributed (the same number for each theory).

NB: significant and robust empirical success (as described in chapter 2) is most relevant here, and throughout this chapter. For brevity, "significant and robust" is omitted in statements of the three conditions for normativity.

Because of the similarities between dissent and consensus, it makes sense to ask whether the normative account used for dissent can also apply to consensus. A further reason for suggesting this is that consensus can be viewed as a limiting case of dissent—when the amount of dissent approaches zero. Consensus is typically not all encompassing, anyway, since usually some dissent remains. The decision whether or not to call a state of affairs "consensus" or "dissent" is to some extent arbitrary.

This chapter will extend the normative account of dissent given in chapter 5 to cases of consensus. This extended account will now be called *social empiricism.* Dissent is treated as the prototype state and consensus as a special case of dissent. The states of dissent and consensus are matters of degree, with continuity between dissent and consensus. Since neither dissent nor consensus are primary goals of science,[1] it may be expected that sometimes more or less dissent, and sometimes more or less consensus, will be conducive to the primary goals, which are (as discussed in chapters 2 and 3) empirical success and truth.

Let's consider the conditions (1), (2) and (3), applied in the special case when dissent approaches zero (i.e., consensus). According to (1), forming consensus is appropriate only if *one* theory has *all* the available empirical successes. Otherwise, other theories have empirical successes of their own, and should be pursued. Moreover, forming consensus is appropriate only if all scientists who are producing empirical successes are doing so with the consensus theory. Otherwise, consensus will, practically speaking, result in a loss of productive engagement with the domain as scientists lose the theories in which they had empirical success. That is, consensus is appropriate only when scientists manage to reproduce their earlier empirical successes, which may not have been with the consensus theory, with the consensus theory. In this condition, which is a special case of (2), empirical decision vectors such as salience and availability will all be in favor of the consensus theory. Finally, in coming to consensus it may be expected that a previously equal distribution of non-empirical decision vectors (as is required in a normatively appropriate dissent) will disappear, as more and more scientists work with the consensus theory. This is normatively acceptable, since the other theories have no extra empirical success. So, in the special case of coming to consensus:

1′. One theory comes to have all the empirical successes available in a domain of inquiry.

2′. This same theory comes to have all of the empirical decision vectors, since all scientists working productively (with empirical success) are working within the one theory.

3′. Any distribution of non-empirical decision vectors is OK, but typically more will develop, over time, on the consensus theory, as the old theories fade away. During dissent, and thus in the early stages of consensus formation, the above requirement of equal distribution of non-empirical decision vectors holds.

Note that, as a consequence of these conditions, empirical decision vectors select the theory on which there is consensus. This means that the empirical successes underlying them cause the choice of consensus theory. Non-empirical decision vectors stay balanced until later. Both empirical and non-empirical decision vectors are typically *necessary* for consensus formation, whether normatively correct or not.

Note also that it is appropriate to form consensus only in the extreme case that one theory has *all* the empirical successes. Traditional epistemologies of science (e.g., those of Lakatos and Laudan) call for consensus when one theory is *significantly superior* to another in, e.g., problem solving success, which is a much weaker condition.[2]

As mentioned in chapter 6, consensus, even when it occurs, is not the end of the story. Frequently, as anomalies develop, new theories are proposed, and new empirical successes are produced, it becomes appropriate to revert to dissent. Dissolution of consensus is appropriate under the following conditions (which are special cases of (1), (2) and (3)):

1″. A new theory has empirical success that is not produced by the consensus theory.[3] (So, the new theory deserves attention.)

2″. Empirical decision vectors come to be equitably distributed.

3″. Non-empirical decision vectors come to be equally distributed.

Again, empirical decision vectors lead the way and have a necessary causal role in producing normatively appropriate dissent, although they are not typically sufficient. Non-empirical decision vectors, as they come to be equally distributed, can be indispensable.

Social empiricism uses the *same* normative framework for assessing dissent, consensus, and the dissolution of consensus. Conditions (1), (2) and (3) are stated in their most general form for dissent. Then the question is asked, under what conditions is it appropriate to have only one theory under development (i.e., consensus)? According to (1), (2) and (3), this will be appropriate under the conditions (1′), (2′), and (3′) mentioned above for formation of consensus. Similarly, (1″), (2″) and (3″) state the specific conditions, derived from (1), (2) and (3), under which dissolution of consensus is appropriate.

Social empiricism is social because what matters, normatively speaking, is the distribution of empirical and non-empirical decision vectors across a community of investigators. Normative judgments are not made of the thoughts and decisions of individual scientists. Social empiricism is empiricist because of the emphasis on empirical success, which was developed in chapters 2 and 3.

Social empiricism requires both more (in some ways) and less (in other ways) of scientific communities than traditional, more individualist, accounts of scientific rationality. Social empiricism is also *more social* than other social epistemologies currently under consideration. I'll expand both of these remarks at the end of this chapter and in chapter 8. At this point, it is time to turn to case studies, and show that social empiricism offers a useful assessment of a variety of case studies of coming to consensus and dissolution of consensus. I begin with the cases in chapter 6.

2 Consensus on Plate Tectonics

Chapter 6 showed that consensus on plate tectonics took place in a piecemeal and socially distributed manner. Between 1958 and 1970, the geological community moved, in subgroups mostly aligned with sub-disciplines, from a protracted pattern of dissent (with roughly equal distribution of non-empirical decision vectors and equitable distribution of empirical decision vectors) to consensus on plate tectonics. Continental paleomagnetists were convinced first, followed by oceanic paleomagnetists, oceanographers, seismologists, stratigraphers and finally continental geologists. Both empirical and non-empirical decision vectors were important. Empirical decision vectors such as salience and availability of particular observations to particular individuals (who made the observations themselves), laboratories (e.g., the personnel at Lamont), conference participants and sub-disciplines were critical for decision making. For the most part, empirical decision vectors inclined decision processes to plate tectonics. Eventually, at the end of the plate tectonics revolution, all empirical decision vectors were in favor of plate tectonics: apparently conflicting data were reinvestigated and reconciled. Non-empirical decision vectors such as pride and belief perseverance acted against plate tectonics, and non-empirical decision vectors such as peer influence (e.g., the positive support of Pitman), peer pressure (later in the plate tectonics revolution) and the influence of prestige (e.g., the prestige of results coming out of Lamont, Scripps and Cambridge) acted in favor of plate tectonics. The non-empirical decision vectors stayed roughly equally distributed until the end of the revolution.

There was no crucial experiment or set of observations—either logically or historically crucial—that brought about consensus in the geological community. Consensus was complete when plate tectonics had all the empirical successes, and a process in which there was equal distribution of non-empirical decision vectors and equitable distribution of empirical decision vectors brought it about. It is, moreover, reasonable to say that the empirical successes of plate tectonics, acting through the empirical decision vectors, were responsible for the choice of consensus theory. Without them, consensus would not have occurred, or it would have occurred, inappropriately, on another theory. Both empirical and non-empirical decision vectors, however, were necessary for the production of a (any) consensus.

Consensus on plate tectonics therefore has the characteristics of a normatively appropriate consensus.

3 Consensus on the "Central Dogma"

Consensus on the "central dogma" also took place in a piecemeal and socially distributed manner. Chapter 6 showed the variety of decision vectors causing consensus: empirical decision vectors connected to the empirical successes of molecular genetics, reductionist metaphysics consisting in a preference for reducing biology to physics and to showing that biology is likewise governed by simple universal laws (this is a non-empirical decision vector), the physical science training of new geneticists leading to ignorance and bias against traditional genetic techniques (again, a non-empirical decision vector), and sexism and anti-Semitism adding to the bias against work done outside the nuclear monopoly (more non-empirical decision vectors). The imbalance of non-empirical decision vectors that began early in the history of genetics (as shown in chapter 5) was continued, with more non-empirical decision vectors in favor of the nuclear monopoly, in mid-century.

Chapter 6 also showed that there were important empirical successes that did not fit into the "central dogma": the work of Sonneborn (on cytoplasmic inheritance and the inheritance of supramolecular structures), McClintock (on genetic interactions) and Sager (cytoplasmic inheritance). The consensus on the "central dogma" caused these successes to be dismissed, delaying important work in molecular genetics. Other empirical successes (e.g., the work of Ephrussi and Nanney) were oversimplified and presented to fit into the "central dogma" narrative. Consensus was not on the theory with all the empirical successes: in fact no theory had all the empirical successes. According to social empiricism, consensus should not have taken place.

Thus, the conditions for normatively appropriate consensus were not met in the 1950s consensus in genetics. Consensus took place on a theory—the "central dogma" that did not have all the empirical successes. Empirical decision vectors were not all with the consensus theory. And non-empirical decision vectors were unequally distributed—even more unequally than they had been during the time of dominance of Mendelian genetics. This consensus was brought about by the unequal distribution of non-empirical decision vectors.

4 Consensus on the Variability Hypothesis

Stephanie Shields (1982) has documented the reception of Havelock Ellis' work on sex differences. Ellis's 1894 book, *Man and Woman,* developed an idea of Darwin's (especially in *The Descent of Man*) that males show greater genetic variability than females. According to this idea, males lead evolutionary change because of their greater variability. Although few differences in variability were in fact found (Pearson, who publicly criticized Ellis' work and even gave evidence to the contrary pointed this out), there was broad consensus on its truth, especially for mental abilities, at the turn of the century.[4] Males were deemed far more likely to be geniuses (as well as more likely to be idiots, but this point was rarely dwelt on).

The variability hypothesis was welcomed as both an explanation and as a justification for women's inferior status (Shields 1982, p. 198). It also resonated with a belief in biological determinism, then popular in the eugenics and genetics communities. Through Thorndike (who refined the statistical formulation of the hypothesis), it influenced educational psychology and testing in the United States at the beginning of the century, and supported placing and maintaining limits on educational and professional opportunities for women. It was reasoned that since women cannot not succeed to the degree that some men can, resources should not be 'wasted' on them.

Even though well-respected educational psychologists (mostly women such as Helen Bradford Thompson (Wooley) and Leta Stetter Hollingworth, but also a man: G. W. Fraiser in a 1919 study that found no differences in variability) criticized the hypothesis, giving both contrary data and alternative environmental explanations of the existing data, broad consensus on the variability hypothesis continued. Much had to happen—scientifically, politically and socially—before this consensus was dismantled.

What went wrong in this episode, according to social empiricism, was not the presence of non-empirical decision vectors *per se,* i.e., not the *presence* of sexist or determinist assumptions, but the *unequal distribution* of non-empirical decision vectors. On the side of the variability hypothesis were coherence with Darwin's ideas, sexism, genetic determinism and the authority of proponents. Non-empirical decision vectors for the other side consisted in Dewey's intellectual influence on Wooley and perhaps some early feminism. Non-empirical decision vectors against the other side consisted in sexism against Wooley and Hollingworth, and their relative lack of prestige (in part because they were junior). This is an aggregate (summing linearly) of 4 decision vectors for the variability hypothesis and none against. Consensus took place on a theory that did not have all the empirical successes, and it is reasonable to identify the cause of this normatively inappropriate consensus in the unequal distribution of non-empirical decision vectors.

This case is like the previous case (consensus on the "central dogma") in that consensus is on a theory that does not have all the empirical successes, and the consensus is brought about by an unequal distribution of non-empirical decision vectors. It differs from the previous case only in degree—there is more inequality of non-empirical decision vectors, and less empirical success of the consensus theory. The next cases will show that it even happens that consensus forms on a theory with *no* empirical success, again because of unequal distribution of non-empirical decision vectors.

5 Consensus on a Surgical Practice[5]

The extracranial-intercranial (EC-IC) bypass operation was first performed in 1967 by Gazi Yasargil, in Switzerland. It consists in taking arteries that feed the scalp and suturing them to arteries that feed the brain. Since the blood vessels feeding the scalp rarely develop atherosclerosis, it was reasoned that the sutured arteries would increase blood flow to the brain, and thus help prevent strokes. Strokes are frequently caused by blockages in atherosclerotic blood vessels to the brain. Right away, surgeons embraced the rationale of the EC-IC bypass and started offering it to patients. Even before Yasargil published data from the 84 bypasses he performed over the next ten years, several thousand EC-IC bypasses had been performed.

Yasargil's data looked good. Only 3 of the 84 patients he operated on had further strokes. Using retrospective data, Yasargil reasoned that the predicted number of strokes was 40. This result solidified a consensus on the use of the EC-IC bypass; it became standard treatment for stroke patients and was covered by health insurance.

The EC-IC bypass required special training in sophisticated microsurgery and was very lucrative. As Vertosick writes, "The operation became the darling of the neurosurgical community" (1998, p. 108). Yet Yasargil's data were retrospective, and did not take into account other recent changes in treatment for stroke patients that might have been responsible for the decreased mortality. A few opponents of the EC-IC bypass urged further investigation, and in 1977 the National Institute of Neurological and Communicative Disorders and Stroke (NINCDS) funded a large prospective study of the EC-IC bypass.

In 1985 the results were announced: patients who received EC-IC bypass surgery did, if anything, a little worse than patients who did not receive the surgery (all patients received daily aspirin therapy, which was already proven to decrease the risk of stroke and which was in wide use). The results were announced by one of the major investigators—Sydney Peerless—at the 35th annual meeting of the Congress of Neurological Surgeons. There was a "brief and acrimonious debate . . . [before] . . . the bypass era finally ended" (Vertosick 1998, p. 108). Vertosick notes that confidence in the bypass operation did not die until insurance companies began denying coverage: "many surgeons still believed in it, but not enough to do it for free" (1998, p. 108).

In this case, consensus formed on a theory with *no* empirical success. Of course, it was thought to have empirical success at the time, and that was one cause of consensus. But consensus was well under way before any data was presented. Reasoning about blood flow to the brain,[6] the allure of an impressive microsurgical procedure and financial incentives were three non-empirical decision vectors influencing the consensus. After a while, personal investment and pride in the technique also played a role. Indeed, until these decision vectors decreased in force (until the financial incentives disappeared) the consensus was not fully dismantled. Probably, the prestige of the NINCDS study and its investigators also contributed to the dismantling of consensus, along with empirical decision vectors such as the salience and availability of data presented at the annual conference of neurological surgeons.

Vertosick does not mention any non-empirical decision vectors on the other side of the debate, and the only one I can think of is conservatism. Typically, some surgeons oppose new techniques because they do not want to change their practices and learn something new. Non-empirical decision vectors were thus unequally distributed (4:1), and the consensus was normatively incorrect.

Vertosick's normative recommendation is that surgical techniques go through the same series of trials (Phase I, Phase II and Phase III) that the FDA requires for pharmaceuticals. I agree with this—this is a good way to ensure that the theory or practice has robust empirical success—but I add the further conditions that empirical decision vectors be equitably distributed and non-empirical decision vectors be equally distributed. Only in that way can we be confident that dissent and consensus form for the right reasons or causes, rather than for epistemically arbitrary ones.

This kind of case has occurred frequently in surgery (the practices of routine shaving before surgery and episiotomy in obstetrics are similar examples that have recently been challenged). Surgery has not been held to the standards of evidence-based medicine (clinical trials, etc.) that internal medicine, for example, now takes for granted.

6 Consensus on the Ovulation Theory of Menstruation[7]

During the early nineteenth century, physicians reasoned by analogy from animals to humans, and equated animal estrus with human menstruation. Since animals ovulate during the period of estrus, it was reasoned that humans ovulate during menstruation. Couples trying to conceive were advised to have intercourse immediately after the end of menstruation.

The ovulation theory of menstruation had little or no empirical success. The consensus was brought about through analogical reasoning, and then reinforced by the usual mechanisms of authority and belief perseverance. The consensus was dismantled in the late nineteenth century, after an accumulation of negative data.

In this case, as with the previous case, non-empirical decision vectors were not equally distributed, and they caused consensus on a theory with no empirical success. In this case, there was not even a claim to have empirical success.

7 Consensus on the Copenhagen Interpretation of Quantum Mechanics

Cushing (1994) argues that consensus in 1927 on the so-called "Copenhagen Interpretation of Quantum Mechanics" was an historical accident rather than a scientific advance. There was an alternative to quantum indeterminacy, even at that time. Causal theories were available that are more continuous, metaphysically speaking, with Newtonian mechanics than the Copenhagen Interpretation is.[8] According to Cushing, the debate was settled in favor of the Copenhagen Interpretation because of accidental factors such as the organization of the physics community around the authority of Copenhagen physicists, the intellectual climate in Germany in the interwar period (the Forman thesis) and the unnecessary weakness of de Broglie's attempts to develop a causal theory. When, much later, a satisfactory causal theory was put forward by David Bohm, it was dismissed on religious grounds (e.g., Pauli) or metaphysical grounds (e.g., Einstein who defended locality and Heisenberg physicists who had a metaphysics of subjectivity), by anti-communist feeling (communists favored causal theories) and by conservative as well as anti-metaphysical opinions among physicists, who wanted the matter of interpretation to stay settled. Cushing also argues (this is controversial) that the two theories are empirically equivalent.

According to social empiricism, consensus on the Copenhagen Interpretation was *not* normatively correct. This is because non-empirical decision vectors were so unequally distributed from the beginning to the end of consensus formation. If Cushing is correct that Copenhagen and causal theories are empirically equivalent, then the consensus was on a theory with all the available empirical successes (the causal theory offers no more) and no scientific harm was done (no empirical successes were lost). Moreover, empirical decision vectors came to be all on the side of the Copenhagen theory. Consensus on the Copenhagen Interpretation was epistemically *lucky*, rather than epistemically neutral,[9] because empirical decision vectors of the Copenhagen Interpretation were not responsible for producing consensus. Cushing argues, plausibly, that the non-empirical decision vectors caused this consensus.

Empirical equivalence of historically competing theories is rare. If Cushing's account is correct, this is a case of empirical equivalence of two theories that are not retrospectively constructed formal variants of one another but in fact are metaphysically far apart and contemporaneous competitors. (Formal variant cases of underdetermination are common, and have been produced by some philosophers of science, notably Quine and Van Fraassen.) Cushing suggests that there may be empirical differences in future work on these theories; at this time, however, that is unclear. Right now, Bohm's theory has no more value to quantum physics (value measured in terms of novel empirical success) than any alternative, untested theory.

Cushing's interest in Bohm's theory is metaphysical: he has an interest in seeing whether the successor to Newtonian mechanics can be metaphysically more continuous with it, and thus metaphysically more consistent with intuitions about determinism and objectivity that are deeply rooted in our intellectual culture. I have not included these intuitions in my discussion of scientific goals, as I see no evidence that they are intrinsically scientific goals (I treat them as non-empirical decision vectors).

Cushing's account is somewhat controversial. Several physicists and historians of physics (e.g., Sam Schweber, Allan Franklin) disagree with his account, claiming that Copenhagen interpretations are, in practice, the only ones that can be successfully used in experimental physics. So, they claim, formal observational equivalence is not the same thing as equal empirical success, and Bohmian physics lacks the latter. (Cushing has retorted[10] that these judgments are biased by preference for a familiar way of doing things—the Copenhagen way.) If these critics of Cushing are correct—and I take no sides here—consensus on the Copenhagen interpretation was progressive, but luckily so, since the cause of consensus was unequal distribution of non-empirical decision vectors. In conclusion, whichever way this case is analyzed, the physics community made a normatively incorrect decision.

The cases so far in this chapter are all examples of coming to consensus. According to social empiricism, however, consensus is not the *telos* of scientific research. In fact, as I have already argued, consensus is normatively appropriate only in the rare case that one theory has *all* the empirical successes. Yet, if this chapter ended here, it might leave the impression that, while I have offered new descriptive and normative accounts of consensus, I agree with the majority of historians, sociologists and philosophers of science who tell narratives of scientific change ending with consensus, and who view consensus in itself as the mark of scientific progress. As already argued in chapters 2 and 3, scientific progress is associated with empirical successes and truth in the theory (or theories) rather than with consensus. So this chapter, which gives a normative account of consensus, should consider not only the formation of consensus but also the dissolution of consensus. I turn to such cases next. The normative framework for assessing these cases was given at the beginning of this chapter.

8 Dissolution of Consensus: Cold Fusion[11]

Before March 1989, it was generally accepted that cold fusion is not a viable energy source. Stanley Pons and Martin Fleischmann announced at that time that they could produce excess energy in electrolytic cells, and that the explanation of this result is cold fusion. This announcement generated much interest, not only from the media, from industry and from government, but also from other physical scientists. Immediately, the consensus that cold fusion would not work broke down, and many scientists took sides, sometimes attempting to replicate or refute the experimental work. Dissent was at a peak in May 1989, and continued for several months. During the period of dissent, the success or failure of the experimental work was unclear. Details of the experimental set-up came out slowly (Pons and Fleischmann did not give a full description for a year), and there was debate about the proper controls and proper analysis of the experiments.

Non-empirical decision vectors played an important role in creating dissent. Pons and Fleischmann both had good reputations, in particular for taking scientific risks that paid off.[12] Pons and Fleischmann were electrochemists, and chemists were far more likely to accept their work than physicists. Pride in one's own discipline is a powerful force; moreover, physicists were already invested in their claims that only hot fusion would work. Cold fusion promised a victory for "little science" over "big science": the apparatus was relatively cheap, quick and easy to set up. Cold fusion also promised a victory for scientists at a marginal institution (University of Utah) over scientists at prestigious science institutions such as MIT and Cal Tech, and this promise appealed to many who generally champion the success of the underdog. Market interests supported work in cold fusion. Some physicists—notably Steve Koonin at Cal Tech—argued, influentially, that significant cold fusion was physically impossible. Other physicists—notably Peter Hagelstein at MIT—argued, less influentially (probably because physicists were already generally unsympathetic to cold fusion), that significant cold fusion was indeed possible in the special conditions of the palladium electrodes.

So, on the side of cold fusion were five non-empirical decision vectors: the reputation of Pons and Fleischmann, pride among chemists, support of "little science" over "big science," support of science from non-prestigious institutions and market interests. Against cold fusion were five non-empirical decision vectors: conservatism and belief perseverance on the part of physicists, belief that physics rather than chemistry was to be believed, support of "big science" (e.g., among hot fusion researchers), preference for results coming from prestigious institutions and Koonin's influential theoretical reflections. This is a pretty good balance of non-empirical decision vectors.

How about the empirical decision vectors? For there to be empirical decision vectors, there has to be some empirical success anchoring the vectors. But this was precisely the area of dispute. It was not possible to tell, for months, whether or not excess heat was produced in the electrolytic cells. Pons and Fleischmann's track record made empirical success seem likelier. Eventually, however, it became clear that there was no excess heat. There was no experimental success and thus there were no empirical decision vectors. To be sure, during the period of dissent, experimenters were influenced by the salience and availability of data from their own laboratories—if excess heat seemed to be produced, they supported cold fusion, and if it seemed not to be produced, they opposed it. These influences, however, turned out not to be empirical decision vectors because they were not associated with actual, robust, empirical success.

(Actually, they can be viewed as non-empirical decision vectors; this would have the result of adding one non-empirical decision vector to each side of the debate, which would not affect the normative assessment above.)

According to social empiricism, dissent is appropriate over those theories that have empirical success. Cold fusion, it turned out, had no empirical success, so, in fact, the dissent was normatively *inappropriate.* According to social empiricism, it would have been best to stay with the prior consensus position that only hot fusion is a significant energy source.

Is this a fair criticism? After all, one of the tenets of naturalized epistemology has been "ought implies can," and no one could have known in March 1989 whether or not the purported cold fusion phenomena were robust. Careful attention to what happened in March 1989, however, shows that investigators really could have done better. Cold fusion was announced to the world, not because Pons and Fleischmann were ready to go public with their apparent empirical success, but because they, and their institution, feared being scooped by Steven Jones at Brigham Young and could not reach an appropriate understanding with him. Moreover, experimental details were not given because Pons and Fleischmann feared losing the credit and the financial advantage of being first with a new important technology. At least the experiment was relatively easy, quick and cheap to perform, so dissent did not last too long.

I am not suggesting that Pons, Fleischmann and Jones should not have worked on cold fusion at all. I am claiming that it was inappropriate to cause dissent in the scientific community before there was any empirical success on the side of cold fusion.

I disagree with Collins and Pinch (1993, p. 72) who argue that in the case of cold fusion, experiments were not capable of settling the issue. As argued in chapter 2, experimental success—or failure—can take a while to emerge from noise.

Moreover, I disagree with the authors of several recent books on cold fusion (e.g., Huizenga 1992, Taubes 1993), who argue that cold fusion was "bad science" and the "scientific fiasco of the century" because of the role of media attention, funding politics, secrecy, etc. According to social empiricism, it was not a "fiasco": the presence of non-empirical decision vectors such as media attention, industry funding, etc., were not a problem in themselves and, indeed, are frequently present in successful science. Moreover, the non-empirical decision vectors were about equally present on both sides (Taubes, in particular, focuses only on those non-empirical decision vectors on the side of cold fusion). The attention paid to cold fusion was a reasonable mistake, based on the reasonable judgment that Pons and Fleischmann probably had produced some empirical success. Perhaps taking more care that there is empirical success before disseminating results could lessen such mistakes, as I suggested above.

9 Dissolution of Consensus: Treatment of Peptic Ulcers[13]

From the beginning of the century until the mid-1980s there was consensus that the usual cause of peptic ulcers is excess gastric acid, produced in reaction to stress or a poor diet. It was thought that excess acid erodes the lining of the stomach and duodenum, eventually causing an ulcer. The condition was usually chronic, and treatment consisted of dietary modification, antacid consumption, stress reducing techniques and, since the 1970s, acid blocking drugs such as Tagamet and Zantac. Treatment alleviated the symptoms and often healed the ulcers, but ulcers frequently returned. Treating ulcers was a lucrative business—around $800 per year for medications, plus surgical bills of over $17,000 if the ulcer hemorrhaged.[14] Acid blocking drugs were the most lucrative pharmaceuticals in the world and ulcer surgery among the most frequent operations.[15] Peptic ulcers affected about 10% of the population in the developed world.[16]

In the late 1970s and early 1980s, Barry Marshall and Robin Warren discovered slow-growing bacteria, which they called *Helicobacter pylori,* in the stomach, especially close to areas of inflammation. Marshall, a gastroenterology resident, was asked by his department chair to assist Warren, a pathologist, in his research. Soon, they found that there was a strong correlation between the presence of the bacteria, gastritis, and peptic ulcers. They hypothesized a causal connection, and suggested that removal of the bacteria, with antibiotics, would cure both gastritis and ulcers. Marshall himself believed that H. pylori was the cause of peptic ulcers by 1983.

In 1985, Marshall attempted to show that H. pylori could cause gastritis by ingesting the organisms himself (his experiments to induce infection in animals had failed). He developed gastritis, and an H. pylori infection (shown by endoscopy) and recovered after taking antibiotics. He did not develop a peptic ulcer. His self-experiment did attract attention, in the tradition of macho experimental research: a very salient empirical success.

The next experiment, performed from 1985 to 1987 in the local Royal Perth and Freemantle Hospitals, showed that antibiotics prevented relapse in peptic ulcer disease. The results were published in 1988.

Paul Thagard (1999) has shown that early reception of Warren and Marshall's results was negative among gastroenterologists, and positive from infectious disease specialists. Gastroenterologists had established procedures for ulcer treatment, which they were reluctant to change. Infectious disease specialists were more likely to take the data seriously, because it was more salient to them. Warren and Marshall took advantage of this, by presenting their work at an international microbiology conference in 1983, which resulted in much interest in H. pylori.

By 1989, there were well over 100 papers published on ulcers and H. pylori (Thagard 1999). Clinical practice was practically unchanged, and pharmaceutical companies pushed acid blockers, as usual. Marshall worried that nothing would happen until there was a new generation of gastroenterologists (Thagard 1999).

If we look at the distribution of decision vectors at that time, Marshall's frustration is seen as justified. Although the empirical decision vectors were equitably distributed (the results were salient to Warren and Marshall, their close collaborators, and to microbiologists), the non-empirical decision vectors were unequally distributed. Conservatism of gastroenterologists, the low status of Marshall (a young unknown from Australia who was viewed by many as "crazy") and the apathy of drug companies[17] were all non-empirical decision vectors that worked against a fair reception of the H. pylori hypothesis. Thus there were three non-empirical decision vectors against the H. pylori hypothesis, and none for it.

Thus, the prevailing consensus that excess acid was the cause of ulcers did break down, but the result was a less wide distribution of belief than would have been desirable. In fact, clinical practice barely changed at all before 1990. The situation changed after 1990. Drug company patents on acid blockers began to expire. Faced with drastically reduced profits, pharmaceutical companies lobbied to make acid blockers available over the counter (and thus more heavily—casually—used, for indigestion) and looked for new products to develop. They developed antibiotic-antacid combinations specifically for use in ulcer patients, more lucrative than plain old antibiotics. Moreover, gradually, prestigious gastroenterologists began to become convinced of the role of H. pylori in causing ulcers, and they used their influence on other gastroenterologists. Non-empirical decision vectors began to balance.

To be sure, the continuing empirical successes of the H. pylori hypothesis played a huge role in change of belief. There were important clinical trials, sponsored by NIH, in the early 1990s. Perhaps the empirical successes alone—via empirical decision vectors—could have caused a new consensus on the H. pylori hypothesis. I am pointing out that it would probably have taken longer without the shift in non-empirical decision vectors around 1990, and it probably would have taken less time if the non-empirical decision vectors had been better distributed earlier.

The NIH consensus conference in 1994 took place after all the action was over—hence the conclusion of the conference was both clear and epistemically uninteresting (usually conclusions at these conferences are more qualified). The conference probably had little effect on medical practice.[18] It informed some of the uninformed medical practitioners (fewer than the NIH hoped) and added prestige and authority to the conclusions (non-empirical decision vectors reinforcing the consensus).

10 A More Social Epistemology

Consensus and breakdown of consensus, when normative, are *led* by changes in empirical decision vectors rather than by changes in non-empirical decision vectors. Changes in non-empirical decision vectors can be important for scientific progress, but they should not be the initiators of overall change in the distribution of research effort. The requirements (1) to (3) ensure this.

Social empiricism evaluates aggregate decisions on a continuum between ideally normative, in which there is a perfect distribution of decision vectors over empirically successful theories, and completely normatively incorrect, in which theories with empirical success are completely ignored and theories without empirical success are championed. Most of the time, the distribution of research effort is somewhere between these extremes. This leaves room for normative suggestions for improvement.

Social empiricism demands *less* than most epistemologies of science. In particular, it does not require individual scientists to make overall impartial assessments. Most traditional epistemologies of science, and even most social epistemologies of science, require that at some stage in scientific reasoning (typically, consensus) individual scientists make overall impartial assessments of the merits of competing scientific theories. Social empiricism does not require this. What matters is not how individual scientists reason—*it's* not *the thought that counts*—but what the aggregate community of scientists does. It is more realistic to expect normative suggestions to yield improved distributions of research effort than to expect them to correct individual scientists of their partiality. I will say more about this relation between social empiricism and naturalism in chapter 8.

Social empiricism also demands *more* than most epistemologies of science. It requires a social level of evaluations and recommendations. It is no longer enough for individual scientists to strive to be perfectly rational, reasonably Bayesian or properly cognitive (to give some examples of contemporary individualist epistemologies). It is not even enough for individual scientists to check and correct one another's work (a common prescription of most social epistemologists). Rather, and instead, what is required is evaluation and change at a systems level. This is the topic of chapter 8.

8

Epistemic Fairness

The common problem, yours, mine, every one's
Is—not to fancy what were fair in life
Provided it could be, —but, finding first
What may be, then find how to make it fair.
Up to our means: a very different thing!
Robert Browning, "Bishop Blougram's Apology"

1 Social Empiricism and Naturalized Epistemology

Naturalism is the most important development in epistemology over the last fifty years.[1] My arguments for social empiricism have been consistently naturalistic in that I appealed to case studies and to the empirical sciences of knowledge (social psychology, cognitive psychology, sociology, etc.) throughout. The purpose of this chapter is to make further comparisons between social empiricism and other contemporary social epistemologies of science against the background of a commitment to naturalized epistemology. This will lead to the development of the concept of epistemic fairness.

Hilary Kornblith (1985, pp. 1, 3) usefully stated the essence of naturalistic epistemology when he wrote that the question "How ought we to arrive at our beliefs?" cannot be answered independently of the question "How do we arrive at our beliefs?." The reason for this is that, on pain of an implausible skepticism, humans should have available adequate methods, heuristics, and strategies for gaining knowledge. Kornblith calls this position "ballpark psychologism," (1985, p. 9) meaning that the ways in which we actually think are at least roughly the ways in which we ought to think. Quine himself went overboard with this view, famously claiming that "epistemology . . . falls into place as a chapter of psychology," (Quine 1969, p. 82) and, at least with this aphorism, losing any normative stance.[2]

Epistemology naturalized, in its early years, had several weaknesses. The first, addressed quickly, was the problem of getting a normative epistemology from a descriptive science of reasoning. Quine, after all, never told us *which* chapter of psychology epistemology is. Many (e.g., Goldman, Giere, Kitcher, Kornblith, Laudan, Thagard) assess reasoning instrumentally, in terms of its effectiveness at reaching goals such as truth and explanatory success. This is particularly appropriate with scientific reasoning, which is readily, and naturalistically, understood as goal driven.[3]

The second weakness of traditional naturalized epistemology is its assumption that *psychology*, rather than other descriptive sciences of reasoning which may operate at the social rather than the individual level, is the science of relevance to epistemology. Over the past ten years, this assumption has been challenged by those philosophers (e.g., Goldman, Giere, Kitcher, Kornblith, Longino, Solomon and Thagard) who have created the sub-discipline of "social epistemology," which treats anthropological, historical, institutional, political and sociological factors as seriously as it does individual psychological factors. Most social epistemologists also think that normative judgments should be, at least partly, at the social level. One of the central goals of this book has been to move beyond the focus on psychology, in both descriptive and normative work.

A third problem with traditional naturalistic epistemology is the assumption of "generality." It is assumed that some processes of reasoning will be found to be generally effective, and others generally ineffective. Work on the disunity of science makes this assumption doubtful.[4] Social empiricism, while a general normative account, does not typically generalize about the effectiveness of any particular reasoning process (or, in my terminology, any particular set of decision vectors).[5]

In practice, naturalistic epistemologists in the Quinean tradition are still strongly influenced by traditional epistemological assumptions and concerns. This is not intended to be, in itself, a criticism. Traditional ideas *might* continue to prove useful in a naturalistic setting. That is, a naturalistic gloss on traditional epistemic findings *might* be all that is required,[6] and, in that case, traditional norms such as "avoid inconsistency," "eliminate motivational bias," and "each person must be able to provide justifications for their theories" would continue to be viewed as ideal standards of inquiry.

In earlier chapters, I showed that traditional naturalistic views such as the "Ivory Soap model" (which claims that the amount of "bias" in good science is insignificant) are inadequate. My strategy was thoroughly naturalistic. I used case studies to show that scientists typically achieve their goals with the aid of "biased" reasoning, and that it is therefore inappropriate, as well as unrealistic, to hold scientists to the traditional standard. Yet, as noted in previous chapters, many naturalistic philosophers continue with the "Ivory Soap model," either in whole or in part. For example, Thagard has continued to claim, at least until the mid-90s, that "hot" cognitive bias has insignificant effects on scientific reasoning. And, for example, Louise Antony (1993, 1995) claims that although humans are biased in some respects (principally, in concept formation and language learning, which are individual cold cognitive biases) they can avoid being biased by motivational, social and ideological factors, and should be held to that standard of freedom from bias. She insists that this is the way to get truth. This is traditional naturalism: very much like traditional logical empiricism, but with a naturalistic gloss.

Most social epistemologists still show a traditional preference for finding significant rationality at the level of the individual scientist. For example, Goldman, Kitcher and Longino all expect social interaction to correct individual errors caused by "biasing factors." As always, I see this as a claim that needs testing. And in the case studies I examined, social interaction does not remove much individual "bias,"[7] nor does it typically need to in order to achieve scientific goals.

Traditional expectations also lie behind assumptions of "generality," such as the idea (in, e.g., Goldman, Hull and Kitcher) that economic incentives can contribute to rationality, but other kinds of "hot" cognition such as wishful thinking and ideological influence do not thus contribute.[8] Some historians of science (e.g., Sapp 1987, Taubes 1993) assume that "biasing factors" internal to the scientific community are generally rational, while those "external" to the scientific community are generally normatively inappropriate.[9] Another "generality" assumption is the assumption in Longino (1990) that individual motivational bias is generally bad for science. Again, I have used a naturalistic approach to argue that, based on case studies, none of these assumptions are correct.

Indeed, the naturalistic and instrumental approach to understanding scientific rationality that I established in chapters 2 and 3 motivated a change in the conceptualization of epistemic terms by chapter 4. In place of the widespread "bias," I use "decision vectors," which is wider in extension (including, for example, theoretical constraints) and appropriately epistemically neutral. Even the conceptual part of epistemology of science cannot be done by conceptual *analysis*.

The scope of epistemology of science is typically broadened on a naturalistic approach. Conceptualizing scientific rationality is no longer enough, even when the conceptualization is achieved naturalistically, with the aid of empirical studies (historical, psychological, sociological, etc.) of science. In a naturalistic normative account, one expects *application* of the concept of scientific rationality to assess particular scientific decisions, and suggestions for improvement of scientific decisions. There is no justification for confining epistemic work to the conceptual part, in a naturalistic echo of traditional analytic epistemology.[10] Moreover, application is not a trivial task that could be done by, e.g., any clear-thinking scientist or philosopher. Social empiricism, in particular, requires identification of all the kinds of decision vectors involved, and some realistic suggestions about altering the distribution of decision vectors to make for a more equitable distribution of research commitment.

These steps (the identification of decision vectors and improvement of their distribution) typically require expertise, and, often, multidisciplinary knowledge and skills. The critical training required to identify presuppositions about gender, for example, is quite different from the psychological training and methods required to detect cognitive bias. And the statistical techniques needed to assess the role of birth order are quite different from scientific and philosophical knowledge of theoretical constraints such as simplicity. Traditional Quinean epistemologists are not equipped with many of these techniques. Nor are they able to address imbalances of decision vectors by making concrete suggestions for creating balancing decision vectors or eliminating unbalancing ones. Without some knowledge of social engineering, such suggestions are unlikely to succeed. "Clear thinking" will not accomplish much in the way of identification or correction of decision vectors; indeed, the point of social empiricism was to cast doubt on both the feasibility and desirability of such a Baconian goal. Naturalization means that epistemology becomes multidisciplinary.

2 Standpoint Epistemologies of Science

Few epistemologies of science devote attention to the identification of decision vectors (or "bias") or to realistic ways for improving scientific decisions (or "addressing bias"). A major exception is standpoint epistemologies of science. It is for this reason that I view them as the deepest challenge to traditional (traditional naturalistic or analytic) epistemology of science.[11]

Standpoint epistemology underlies the moderate and radical positions in feminist philosophy of science.[12] A position misleadingly called "feminist empiricism,"[13] espoused by many feminist scientists, and by philosophers such as Louise Antony, is at the more conservative end. Feminist empiricists treat "sexist bias" like other "biases," i.e., as for the most part important to eliminate, and eliminable by clear thinking.[14] At the radical end, feminist postmodernism claims that there are multiple feminist standpoints, corresponding to the additional complexities created by race, class, nationality, age, disability, etc. I treat feminist postmodernism (e.g., the ideas of Haraway, the recent work of Harding) as a kind of feminist standpoint theory.[15]

Feminist standpoint theorists, notably Harding, Haraway, Keller, Longino and Nelson, argue that there is no neutral perspective from which to identify and eliminate "bias." They embrace the idea of "situated" or "perspectival" inquiry and each argue, in various ways, that normative judgments and recommendations are still possible. Moreover, they argue that some "biases" (decision vectors) are only identifiable from a nonneutral perspective. For example, Haraway claims of political perspectives, "Vision is better from below" (1991, p. 190). Social empiricism is itself a standpoint theory, holding that all knowledge practices are partial and situated, although it is not a feminist standpoint theory in particular. Social empiricism, like standpoint theories, also rejects the idea of an epistemology organized around the traditional concept of "bias," claiming that a normative epistemology of science will incorporate, in a positive role, factors traditionally viewed as negative and "biasing."

Antony and other feminist empiricists are not impressed by the standpoint approach. Antony coined the term "bias paradox" to characterize the conundrum facing feminist epistemologists: if the goal is truth, and all positions are biased, and bias reduces the chances of getting truth, why regard feminist epistemologists as any more credible than the gender-biased scientists they are criticizing?

There are two reasons not to take the "bias paradox" seriously. First, investigators are often biased in some respects and not others. If one wants "unbiased reasoning," one can pick areas in which investigators are unbiased. "Bias" is not a property of an investigator, but, at most, a property of a type of reasoning. Second, and much more important, is the result from previous chapters that "bias" is not generally an epistemic flaw. Decision vectors of various kinds can often contribute to scientific success and truth. Of course, I am not the only one who currently argues this. Early work in social epistemology (e.g., Goldman 1992, Kitcher 1993) suggests that "biasing factors" such as the desire for credit can have an epistemically beneficial effect on the distribution of research effort. And, for example, standpoint feminist philosophers of science (e.g., Keller 1985, Longino 1990) show that ideologies lead to preferences for theories deserving more attention, such as Barbara McClintock's non-hierarchical view of genetic interactions. "Biasing factors" are assessed not for their conformity to some a priori idea of good reasoning, but for their actual outcomes. Social epistemology, and standpoint feminist epistemology, are in this way both naturalistic, and indeed, more naturalistic than the misnamed "feminist empiricism."

In chapter 4, I described various new approaches to understanding the epistemic role of decision vectors. With one exception, I showed in succeeding chapters that these approaches do not go far enough. The exception is the most normatively sophisticated approach—Helen Longino's view. In my view, Longino's view is also the most normatively sophisticated of feminist standpoint epistemologies. Others, such as Haraway,

Harding, Keller and Nelson simply advocate more pluralism, perhaps with special emphasis on politically suppressed approaches. Typically, they claim that this makes critical reflection possible.

In the remainder of this chapter, I will assess Longino's work. Then I compare the goals and positions of feminist standpoint epistemology, more generally, with social empiricism. It turns out that there is an important difference as well as similarities. In conclusion, I introduce a particular idea of epistemic fairness, which captures the essence of social empiricism.

3 Longino's Epistemology of Science

Longino recognizes that traditional canons of scientific method (simplicity, predictive power, etc.) are not sufficient for doing science. Scientific work is underdetermined by those canons. According to Longino, "values," which come from ideologies, fill the gap. Different values lead different scientists to develop and pursue different theories. The result can be a healthy pluralism. Moreover, scientists who do not share, say, the dominant value system, are in an especially advantageous position: they are able to identify and criticize values that are often unconsciously presupposed by others. This is a naturalistic, standpoint feminist claim, since Longino argues the point with a number of cases. For example, she shows that feminist biologists discern and criticize linear hierarchical presuppositions in mainstream work. So far, Longino's views are similar to those of several standpoint feminist philosophers of science (e.g., Haraway, Harding, Keller, Nelson).

Longino (1990) goes further, producing a full normative epistemological account. She argues that ideological factors benefit science when the scientific community is structured so as to be, in her sense, "objective." This requires equality of intellectual authority, public forums for criticism, responsiveness to criticism and shared standards including the standard of empirical adequacy. Longino expects that "individual biases" such as cognitive bias, error, motivational bias, etc., will be eliminated through social criticism and that deeper values, or "ideological bias," while not eliminated (this would be both impossible and undesirable), will be made manifest and treated more democratically.

Longino's account of scientific objectivity is clearly socially, rather than individually, applicable. It is in this way more social than most other social epistemologies, which typically include "social factors" only insofar as they eventually contribute to individual enlightenment (e.g., by correcting error, or through eventual like reasoning of individuals in consensus). It has a certain intuitive plausibility: the requirements of greater intellectual democracy would seem to give all scientists a fairer hearing, thus increasing the chance that good ideas will be considered. Longino also claims to be a naturalist. The important question in this context is: how naturalistically informed (rather than how intuitively plausible) is the account?

A naturalistic justification of Longino's account of scientific objectivity needs to appeal to case studies, or at least to social science theories with evidential support. The case studies should show that requirements for scientific objectivity are both attainable and scientifically desirable. For example, it might be found (in past cases or current observations and experiments) that responsiveness to criticism generally advances scientific work, or that scientific communities with less of a hierarchical structure produce more or better science. Longino does not do this. To be sure, her account of scientific objectivity may have its *source* in reflection on actual cases in which Longino herself (with close colleagues, e.g., Ruth Doell) experienced the frustration of having good, and unrecognized, scientific work. An epistemic remedy to this situation *might* be a more democratic scientific community. But it might not be the only remedy, or the best remedy, or even a feasible remedy. There is already reason for doubt. For example, criticism frequently results in entrenchment of prior positions (due to pride, confirmation bias, etc.). This is clearly not the "response" that Longino has in mind. And, for example, no one has a realistic model for a scientific community in which inequalities in intellectual authority do not play a large role. Naturalism is as important an approach for finding effective ways to correct epistemic problems as it is for identifying them.

Longino's account of scientific objectivity appears to be the product more of logical and rational reflection than of naturalistic investigation. Longino accepts that values and ideologies are proper influences on scientific reasoning because she thinks that scientific reasoning is *incomplete* without them. The first three chapters of *Science and Social Knowledge* argue that contextual values (as opposed to constitutive values such as empirical success and simplicity) are necessary on logical grounds for doing science. Longino's expectation that "individual biases" can be, and should be, eliminated by social criticism is not supported by actual cases. That is, Longino does not justify her distinction between ideological factors (which she glorifies as "values") and individual factors (which she regards, with more traditional epistemologists, as "biases" to be eliminated). Moreover, her requirements of "equality of intellectual authority" and "responsiveness to criticism," while sounding reasonable, are not shown to be feasible or effective by application to particular cases. Because of this, there is no reason for thinking that Longino's normative account is any better than the traditional naturalist ideal (i.e., logical empiricism with a naturalist gloss).

In fairness, Longino explicitly regards her account of scientific objectivity as an *ideal,* which actual cases approach to a greater or lesser degree, resulting in degrees of objectivity (see 1990, p. 80). But even this more modest claim is not supported by cases.

4 Social Empiricism and Feminist Philosophy

Feminist philosophy is a complex, diverse and fuzzy bordered field. General classifications into "feminist" and "mainstream" epistemology and philosophy of science say little about content, although they are sometimes of strategic importance. In this last section, I will focus on content, and explore some common ground between social empiricism and various feminist philosophers. First, I will mention the similarities and one difference between social empiricism and several standpoint feminist philosophies of science. (I find "feminist empiricism" to be so close to the traditional naturalist views already considered that I do not consider it separately.) Then I will describe a more subtle similarity between social empiricism and feminist political philosophy.

Similarities

- In social empiricism, scientific rationality is both socially emergent and socially applicable. While the details differ, several feminist philosophers of science have the same view—notably Donna Haraway, Sandra Harding, Helen Longino and Lynn Nelson. For example, Haraway writes, "[Objectivity arises from] the joining of partial views and halting voices into a collective subject position" (1991, p. 196)

• Helen Longino has a straightforward criterion for feminist philosophies: that gender is not "disappeared" (Longino, 1994). Gender is "disappeared" in an epistemology like Quine's, which merely conceptualizes bias. More naturalistic epistemologies, such as standpoint feminisms and social empiricism, realize that the identification of bias (or decision vectors) requires specific skills. Thus the identification of decision vectors relating to gender needs specifically dedicated effort and should not be "disappeared" into, for example, a general effort to be clear-headed and follow scientific method.

• Social empiricism does not find that the distinction between "reason" and "passion," or between "cold cognition" and "hot cognition" is of epistemic significance. Feminist epistemologies agree, and go further, stressing that this familiar dichotomy is gendered in ways that are both epistemically and politically inappropriate.

• The "gladiator theory of truth" (as it has been described by feminist critics) is the view that inquirers aim for the one true theory of the world, slaying all other competitors. It is widespread among traditional naturalistic philosophers of science,[16] as well as in pre-naturalist (e.g., logical empiricist) views. Social empiricism rejects this without rejecting truth as a principal goal of science: more than one theory can have truth in it (chapter 2).[17] Most feminist philosophers of science also reject the gladiator theory of truth (some also reject all claims to truth, and this is a difference with social empiricism).

• Standpoint feminist epistemology stresses that knowledge is always obtained from some political perspective and that epistemic neutrality is, therefore, impossible. Moreover, particular political perspectives can be epistemically advantageous, allowing investigators to see hidden assumptions in other views. Harding, for example, quotes Haraway (1991, p. 188): "Feminist objectivity means quite simply situated knowledges" and goes on to say, "Standpoint theories argue for 'starting off thought' from the lives of marginalized peoples; . . . [this] will generate illuminating critical questions that do not arise in thought that begins from dominant group lives" (Harding 1993). Longino, Keller, Nelson and others each express similar views. They, and others (e.g., Patricia Hill Collins 1986) give examples of the questions and ideas that have resulted from various marginalized social perspectives. Social empiricism agrees that epistemic neutrality is impossible, unnecessary and often undesirable. Any viable account of objectivity must acknowledge the partiality of scientists. And a good assessment of the rationality of scientific decision making needs to identify decision vectors. Members of marginalized groups often have special abilities, deriving from their social locations, to identify particular kinds of decision vectors.

Difference

• Social empiricism is an epistemology for science. It discerns the epistemic goals naturalistically, by investigating the actual goals of scientific activity. In chapters 2 and 3, I argued that the primary goals here are empirical success and a related goal, truth.

Many feminist epistemologies, on the other hand, often have other or additional goals. Some support inquiries into technologies that might benefit women, such as medical care for women's illnesses. Some support directions of research that explore women's activities, such as woman-the-gatherer theories. And some—notably Haraway, Keller and Longino—favor theories that are supported by an interactive rather than a hierarchical ideology, since hierarchical ideologies lie behind traditional gender relations and other unjust institutions. In general, feminist advocacy is pursued along with the goal of scientific success. In my view, this is not a factual disagreement.[18] Rather, it is a difference in the goals pursued in a particular academic context.[19]

A related point is that social empiricism is interested in *all* sources of partiality, not only those emphasized by feminist philosophers of science who focus on the partiality arising from oppressive political structures and ideological frameworks. Some partiality may, for example, be due to cold cognitive bias (e.g., salience of some data, confirmation bias), and some to hot cognitive bias (e.g., egocentricity, risk aversiveness). Again, I see this not as a factual disagreement but as a difference in emphasis, due to the different overall goals mentioned in the previous paragraph.

There is nothing objectionable, unusual or even particularly feminist about adding non-epistemic goals to an epistemic enterprise. To give another example of non-epistemic goals, the contemporary practice of consensus conferences in medicine is largely driven by practical and political goals such as the desire to improve clinical practice by disseminating universal standards, and the desire to keep decision-making power in the hands of physicians (rather than, e.g., insurance providers).[20]

Standpoint feminism, at its best, has a strong commitment to naturalism.[21] It is this commitment to naturalism that leads to the acknowledgement that all knowledge and knowledge practices are situated. Standpoint feminism emphasizes the ways in which the situatedness is political, and is politically committed to ensuring that politically subjugated situations get epistemic attention. There is no substitute, in standpoint feminism, for the actual epistemic standpoint of those in subjugated positions. That is, there is no "imagining" what a white middle-class woman, or native American, or professional black man, or Asian woman might say. A commitment to naturalism is a commitment to what is and not to what one thinks there might be. In my view, social empiricism is close to standpoint feminism because they both attempt to be thoroughly naturalistic.

5 Conclusions

The simplest way to contrast standpoint feminism with social empiricism is to say that the former is more concerned with political fairness (or justice) and the latter is more concerned with what I will symmetrically call *epistemic fairness*. Note that just as political fairness evokes the social side of ethics, epistemic fairness evokes the social side of epistemology. Conceptualization of bias (or decision vectors) and identification of bias (or decision vectors) are similar in both approaches, although social empiricism discusses more kinds of bias (decision vectors). The difference between the two kinds of approach shows up in the third epistemic task, of remedying poor distribution of bias (decision vectors). Here, standpoint feminism and social empiricism have different goals, that is, different views of what is important to remedy.

Social empiricism is primarily interested in bringing about democratic *science.* This is not the same as a democratic *scientific community,* which may or may not produce epistemic fairness. Scientists could have equality of intellectual authority and still be systematically skewed in favor of a particular theory, for example if they were unduly conservative and risk-aversive. Democratization of a scientific community at most balances the political decision vectors.[22] Many other non-empirical decision vectors need to be equitably distributed in order to achieve epistemic fairness (a democratic science).

A natural question to ask at this point is, who is the normative social empiricist? The traditional view is that the applied epistemologist is a philosopher, able to think clearly, discern bias, and recommend correctives. That is, the traditional view is that producing epistemic justice is armchair work that can be done by sharp-minded individuals. Everything that has been said above about "bias," however, indicates that this idea is naive and non-naturalistic.

The same question arises in matters of political justice. Rawls's *Theory of Justice,* which has dominated the field of mainstream political philosophy for nearly thirty years, has been widely criticized for its central idea, the veil of ignorance. The veil of ignorance is a thought experiment, which places humans, imaginatively, in a situation in which they do not know any particular facts about themselves (social status, personality traits and skills, special relationships, etc.). Impartiality is thought to occur as a result of being able to *imagine* all points of view and compare their desirability, using supposedly universal standards of rationality and reasonableness. This implies that a clear thinking and empathic individual can produce principles of justice on their own, in an armchair. As Susan Moller Okin (1989) has pointed out, Rawls himself shows some failure of imagination: the position of women is almost never considered, whereas the position of other oppressed groups, such as racial minorities and the economically disadvantaged, is regularly considered. Longino might say of this: Rawls "disappears" gender. Young (1990), Sandel (1982) and others argue more generally, and on naturalistic grounds (see, e.g., Young 1990, p. 105), that we cannot imagine all positions in Rawls' abstract way. A related point (argued by, e.g., Young 1990) is that there are no universal standards of rationality and reasonableness, so that even if the veil of ignorance is a worthwhile thought experiment, an individual cannot do it.

Many, especially feminist political philosophers, think that Rawls's position is, for these reasons, untenable. Instead, they suggest that affirming difference and special relationships, rather than abstracting from them, best achieves justice. For example, Iris Young supports difference with new institutions and practices such as affirmative action. Critical legal theory does the same.

I am not going to push the comparison between political justice and epistemic justice too far, since I am not a fan of the analogy game.[23] One similarity is especially obvious: both correct injustice not by utopian claims to be able to remove prejudice or bias, or imaginative claims to be able to think without it, but instead by practical measures to discover and compensate for it.

Who makes the recommendations for bringing about epistemic fairness, if not the "clear thinking philosopher"? No individual is in a position to identify all the decision vectors in a particular scientific controversy. Any individual with normative responsibilities (e.g., a grant officer) can consult experts on various kinds of decision vectors (e.g., social scientists, rhetoricians, social psychologists, gender theorists, etc.), as well as experts on the particulars of a case (e.g., historians, journalists, involved scientists). If the discovery is that there is inequitable distribution of research effort, ideas about balancing decision vectors can be made realistically in consultation with those knowledgeable about social interventions.

The normative social empiricist need not be a professional philosopher or logician. Anyone who is situated so as to be able to both assess and influence the distribution of research effort—grant officers, science policy experts, sometimes department heads, journal editors—can do so as a social empiricist, in consultation with relevant experts on various decision vectors. This is a new locus of epistemic responsibility. Normative suggestions in philosophy of science are typically addressed to the individual working scientists. Social empiricism focuses, instead, on epistemic responsibilities at the level of policy.

Social empiricism is not completely indifferent to the decisions of the individual scientist. One norm that applies at the individual level is: aim for, and work with, theories having empirical success. Social empiricism moves to the social level, though, for many normative assessments, such as assessing comparative worth of various theories when each has empirical success. It does this by discussing distribution of decision vectors. These assessments are, to use some traditional terminology, external to individual awareness: individual scientists are not expected to be able to make them. And distribution of cognitive labor can be normative without *anyone* having made that assessment. As I said in chapter 7, it's *not* the thought that counts.

Social empiricism is conceptually simple. There is nothing mathematically or philosophically challenging in the idea that empirical decision vectors should be equitably distributed and non-empirical decision vectors equally distributed. Difficulties come in identifying decision vectors, and in making realistic recommendations for changing their distribution.

The epistemic advice yielded by social empiricism is not of the kind that *guarantees* success. Social and historical complexity makes attempted manipulations somewhat fallible. There are already built in approximations, such as use of an improper linear model for assessing distribution of decision vectors. This does not mean that recommendations are worthless. We attempt reform in many complex areas of life (e.g., public health, education) so, why not science? And, practically speaking, traditional accounts of scientific method have not even offered this much. Social empiricism gives new epistemic advice, at a social level that has not previously been considered.

Notes

Chapter 1

1. Many sociologists of science did not become such "sociologists of scientific knowledge": they did not become social relativists or constructivists (i.e., members of the SSK movement, called, in its various forms, "the strong programme," "the Edinburgh school," "the Bath school," etc.). They continued work in the tradition of Mertonian sociology of science.

2. Le Grand (1988) is an explicit example of this.

3. Other kinds of rationality include acting in one's self-interest, commonsense reasoning, mathematical/logical reasoning. It is important not to equivocate on the use of the term when exploring one sense, e.g., scientific rationality.

4. Another reading of Feyerabend is possible. Feyerabend may have been saying "*if* you think method is general, then you'll find there is no method at all." That is, Feyerabend can be read as challenging the generality assumption without challenging the idea that there are scientific methods.

5. Van Fraassen's well-known distinction between semantic and syntactic models of scientific theories does not significantly qualify this claim since the "semantic model" still requires a linguistic apparatus.

6. See Solomon 1995a for a fuller exploration of the topic of naturalism and generality.

Chapter 2

1. I take these reflections to be obvious, and of a piece with ordinary discussions about the goals of various enterprises (not only medicine, but also football, stamp collecting, gardening, etc.). Some sociologists of scientific knowledge (e.g., Collins, Woolgar) and some feminist critics of science (e.g., Nelson, Longino) think that anything *can* be a goal or value of science. This discussion, as well as the discussions in chapter 8, counters these views.

2. In this context, "realism" and "antirealism" are used as shorthand for "scientific realism" and "scientific antirealism."

3. In this I agree with Hull (1988), Laudan (1987) and others who also evaluate scientific rationality instrumentally.

4. The difference between empirical and theoretical successes is made on empirical grounds (although it may be an obvious difference, it is not an a priori or "purely conceptual" difference).

5. Longino considers these feminist values because they are democratic values. At root, she thinks feminism is about democracy (1990, 1994).

6. Longino's equivocations—in which she vacillates about the negotiability of particular empirical successes—are not relevant here.

7. Richard Boyd (1980) has argued similarly that methodologies of science are themselves subject to test, and improve along with the general progress of science. Boyd's view of the nature of scientific progress (he argues for convergent realism, approximate truths) is, however, different from mine.

8. Those with a very short lifespan—bacteria and viruses—may be used for experiments on evolutionary change. Paleontological evidence of them is, however, slight at the present time (there is some in Antarctica).

9. I have relied on Bowler (1983) and Bowler (1988) for this account.

10. Early in the twentieth century, with the development of Mendelism, the mutation theory was revived—but this time, the mutations expected were on a smaller scale (one genetic trait rather than a new breeding population overnight).

11. Elsewhere (Solomon 1995c) I have understood this kind of statement as marking a pragmatic turn in naturalistic philosophy of science.

12. As reported by Folkman in a talk to the American Philosophical Society, May 24, 1999.

13. More precisely, I think scope is a theoretical value. Theories with wide scope are not preferred when they are not empirically successful.

Chapter 3

1. Thomas Nickles (1992) has argued that whig history can be good science.

2. Berkeley is usually thought of as the first to systematically develop an antirealist philosophy of science. Antirealism (and related positions—phenomenalism, instrumentalism and conventionalism) became much more widespread about a century ago, with the work of Mach, Duhem and Poincare, perhaps because of developments in quantum mechanics and relativity theory.

3. AI has empirical success when it successfully models psychological processes.

4. Strong social constructivist positions do not have this view about empirical success. These positions were considered and rejected in chapter 2.

5. This need not be read at the general level (truth in general explains success in general), thus avoiding Fine's (1986) objections. Rather, particular truths (correct descriptions of states of affairs) explain particular empirical successes.

6. Psillos (1999) has a defense of realism that is similar, in important ways, to Kitcher's. He also offers a new theory of reference and a new theory of the essential core of a theory. The kinds of criticisms I make of Kitcher's view apply to Psillos, also. Kitcher's work is earlier and better known, so I focus on the details of his account here.

7. Chance: the empirical success of the theory was due to luck. Choice: a theory was chosen for particular empirical successes while others were rejected because they do not have this success; the theory has not been used successfully to make new predictions.

8. Some might argue (e.g., Cartwright 1999) that the meaning of laws is given by their practical use, and therefore Newton's laws were true. This is a dubious semantics: Newton's laws were regarded as universal, for all masses and velocities. Revisionary semantics is just another way to do a whiggish analysis, and in the end no different from (and more confusing than) my analysis in terms of "truth in the theory."

9. As quoted in Bowler 1983, p. 5.

10. An exception was Brown-Sequard's work on the inheritance of induced epilepsy in guinea pigs (Bowler 1983, p. 67), which was an empirical success of Lamarckism (to my knowledge, currently unexplained).

11. Burian 1986 discussed "Lillie's Paradox": in Mendelian genetics, there is no account of cellular differentiation: if chromosomes rule cellular processes, how can very different cells within the same organism be produced and controlled by the same chromosomes?

12. Expansionism was proposed in 1857 by Richard Owen, but it did not get a following until the 1960s.

13. A pragmatic reason for preferring consensus might be to create unity and cooperation in the scientific community. But how much unity and cooperation is needed for scientific success?

14. For example, Kitcher, Thagard, Giere regard times of dissent as valuable only because they promote division of labor, and they argue that division of labor is the most efficient way of discovering which theory is the right one. More will be said about this in chapter 5.

15. The attitudes Fine is speaking of are, actually, deterministic and indeterministic views of the world. I think additional argument is needed to show that these attitudes correspond to scientific realism and antirealism.

Chapter 4

1. The term "rationality" has multiple meanings. Even in epistemological contexts, it is ambiguous, with meanings ranging from "self interest" to "reasonable"

to "conforming to a logic" to "logically truth conducive." I use the term "scientific rationality" to focus this discussion on what is epistemically relevant in philosophy of science.

2. The conflation of reasons and causes here is deliberate. It is assumed that causes of choice sometimes instantiate reasons for choice.

3. A scientist may *believe* that complex interactive theories are more empirically successful than simple hierarchical ones. If so, the scientist is prejudging the issue. What matters here, for classification of decision vectors, is whether or not the scientist is right. And this is an empirical matter.

4. The representativeness heuristic covers much, according to Nisbett and Ross (1980, chapter 6). It underlies both resemblance thinking and analogical inference.

5. Psychologists have for the most part investigated "cognitive bias" in everyday reasoning, rather than scientific reasoning. Although they occasionally use scientists as subjects, they rarely test them in their professional domains. Faust (1984) discusses the existing data on "cognitive bias" in scientific practice. It largely comprises cases of error in medical and psychiatric diagnosis and "confirmation bias" in peer review of scientific work. Faust only speculates, and encourages research, on the presence and implications of "cognitive bias" in other aspects of scientific practice such as theory development, experimental evaluation and choice between competing theories.

6. This is, according to Sulloway, the proximate cause of sibling differences in personality. He has a sociobiological explanation of the ultimate cause, which is irrelevant here.

7. Recent work by Nisbett et al (forthcoming) indicates that the influence of these heuristics is culturally variable.

8. There are other causes of failure to change beliefs under positive and negative undermining. Availability and salience are thought to be the "cold" cognitive causes; there are also "hot" motivational causes such as pride and wish fulfillment.

9. This would require a psychological theory.

Chapter 5

1. In this context, Kitcher claims that while scientists do not disagree with one another, they choose different theories to work on because they are motivated by a desire for credit.

2. I say this because even when scientists start out by simply pursuing a theory, they tend to end up believing it, because of confirmation bias. Thus any initial cause of distribution of research effort tends to be supplemented by causes resulting in eventual disagreement of scientists with one another.

3. Husain Sarkar (1983) also holds this view.

4. Elsewhere (e.g., Kuhn 1962), Kuhn sounds much more cynical about "objectivity," and emphasizes the role of social and motivational factors. His position may have changed since 1962.

5. Kitcher and Goldman argue this, using economic models. The others assume.

6. Laudan and Sarkar and Kuhn might be exceptions to this. This will be discussed in chapter 6.

7. As defined by Kahneman, Slovic and Tversky (1982) and Nisbett and Ross (1980).

8. Orthogenesis is the view that species evolve through linear, internally or metaphysically predetermined stages, from early primitive species to complex higher animals, ultimately humans. Hyatt added to this that sometimes the line of development ends degeneratively, in "racial senility" or extinction.

9. Neo-Lamarckists simply refined their theories, arguing that negative traits were not inherited.

10. Some might argue that consistent theories, for example, are likelier to be true (and choice based on consistency is choice based on a non-empirical decision vector). While this might be so in artificially simplified cases, I don't think that it typically is in real cases. Usually, theories are too complex to be properly checked for logical consistency. Moreover, in practice theories are used in ways that avoid dependence on known inconsistencies.

11. I deliberately do not say, "cancel out," since decision vectors do not, generally, physically cancel out. Some might want to describe them as *logically* canceling out, but such language is misleading in this naturalistic account.

12. An example is the rhetorical effects during the plate tectonics revolution, when data for drift were presented at conferences in an especially salient manner. See the next chapter.

13. Mendelism separates embryology and development from heredity, and leaves the questions of development unanswered. Historically (from the late nineteenth century) inheritance and development were treated together.

14. Even though the eugenics movement began with Galton, who formulated the law of ancestral inheritance (which is pre-Mendelian) and believed in saltative evolution, and even though some eugenicists were neo-Lamarckian, Mendelism became the most attractive theory for eugenicists, for ideological reasons (Mendelism could be used to defend the social status quo).

15. Morgan is an exception to this. He developed Mendelism after years of frustration with embryology.

16. Some might describe the history of genetics as composed of more than one research tradition, dividing non-Mendelian approaches into various different approaches. If this is done, the distribution of non-empirical decision vectors is even worse. I have not done so, however, because the non-Mendelian approaches are closely related to one another.

17. No clear distinction need be assumed between political and ideological factors, and other decision vectors.

18. Expansionism was not yet a theory with any significant evidence or following, although Richard Owen had proposed it in 1857. (Le Grand 1988, pp. 28–29)

19. See Sulloway 1998. Age is also correlated with conservatism.

20. Australian biologists (e.g., Harrison, Nicholls) were even more sympathetic to drift (see Le Grand p. 86–89), but I am confining my discussion to decision vectors among geologists.

21. This is of course a matter of degree. I have yet to find a case where it can be definitively argued that no political or ideological factors played a role.

22. This analysis is of cancer research in the United States only. It is possible that the analysis would turn out differently if work done in other countries was included. Adequate historical work on this is simply not available at this time.

23. Here lies the answer to the frequent question, "What is the status of astrology? Of creationism? Etc." According to social empiricism, when these sciences have robust and significant empirical successes, they are worth pursuing, even when they contradict established theories. Acupuncture may be such an empirically successful science.

Chapter 6

1. Mill (1982 [1859], p. 163) explicitly states that his views in "Of the Liberty of Thought and Discussion" apply to "natural philosophy," i.e., the sciences.

2. Reichenbach (1938, chapter 1) is one source of this distinction, generally popular among logical positivists and logical empiricists. I am grateful to Alan Richardson for this historical reference.

3. I have found only one philosopher of science who explicitly says that the dissent/consensus distinction is the new discovery/justification distinction. This is McMullin (1987).

4. This is a dubious view of evolutionary change as well as of scientific change.

5. A transparent example of this is Kitcher's use of a military analogy in 1993, pp. 203–4.

6. As discussed in chapters 2 and 3, this is because in science truth comes as a result of significant and robust empirical success.

7. As Menard (1986, p. 194) has argued, Wilson's data suggesting that the age of islands is proportional to their distance from a ridge was actually misleading and false.

8. It should be noted that Menard himself was a pioneer in oceanography.

9. Soviet genetics was still dominated by Lysenko, and unproductive, until the mid 1960s (see, e.g., Sapp 1987, p. 165)

10. Of course, consensus is not defined as 100% agreement. Typically, there are dissenters to every view—and this dissent can be productive or unproductive (of

experimental success). There is a continuum between dissent and consensus, and nothing I say depends on precisely where the line is drawn. Chapter 7 will show that the same descriptive and normative accounts apply to both.

11. For example, few ask about differences between Dolly the sheep and Dolly's nuclear parent, and similarities between Dolly and Dolly's cytoplasmic parent.

12. I haven't seen this done in print, but it has come up in conversational responses to my work.

Chapter 7

1. The current emphasis on "consensus conferences," especially in medicine, contributes to a widespread assumption that consensus in itself is good for science. Consensus may be good for health care policy and pedagogy, but there is no evidence that it is good for science. Indeed, the point of the genetics example in the previous chapter was to show that consensus can get in the way of scientific research. Similar examples will be given in this chapter.

2. Contrast the traditional philosophies of science with Feyerabend's comment, "A scientist who is interested in maximal empirical content . . . will accordingly adopt a pluralistic methodology" (1975, p. 47). Feyerabend, however, thinks that it is *never* appropriate to come to consensus.

3. Usually, any empirical success can be produced in ad hoc fashion. But this would not be significant empirical success, in the sense discussed in chapter 2 and required for conditions (1) to (3).

4. There was broad consensus in the USA and the UK. To my knowledge, these were the places where the hypothesis was considered.

5. This account is taken from Frank T. Vertosick Jr (1998).

6. "Reasoning" here is a non-empirical decision vector because it is not connected to any empirical success of this particular theory (even though it is connected to the empirical success of other theories).

7. My account is taken from Marsh and Ronner 1996.

8. These causal theories are more continuous with Newtonian mechanics, although not completely so, since they allow non-local causation.

9. Neutral rather than progressive, because according to social empiricism there is nothing progressive about forming consensus unless only one theory has all the empirical successes.

10. E-mail correspondence.

11. Sources are Taubes 1993, Collins and Pinch 1993, Huizenga 1992.

12. Scientific reputation has several causes and features. I classify it as a non-empirical decision vector because unearned authority is typically at least part of scientific reputation. Often, there is also earned authority. I view that as the source of good reason to trust a scientist's judgment that he or she has produced scientific

success. That is not in itself an empirical decision vector *for the theory under consideration,* since the trust derives from proven empirical successes in other domains.

13. Sources are Thagard (1999), *Scientific American* (February 1996), *Economist* (March 5, 1994), Barry Marshall's website (see footnote 17 below).

14. See *The Economist,* March 5 1994, p. 110, for a calculation of the costs of treating ulcers.

15. *Scientific American,* February 1996 (pp. 104–107), article by Martin Blaser.

16. Thagard (1999).

17. To call it "apathy" is to be generous. This is the word Marshall uses (http://www.helico.com/cgi-bin/HyperNews/get/diagnosis/6/6/3.html). Patents for acid blockers did not start to expire until 1990, and then things began to change.

18. Studies have shown that, in general, consensus conferences have less than the desired effect on practice. See Ferguson 1993.

Chapter 8

1. It is arguable that naturalism is also an old idea, eclipsed by Frege and then the schools of analytic philosophy and logical empiricism. Recent naturalism is in any case particularly significant because of concurrent developments in sciences relevant to epistemology.

2. Quine made up for this elsewhere, with his espousal of a traditional view of scientific method, characterized by the goals of predictive power and simplicity (e.g., Quine 1960 and later work).

3. This was argued in chapter 2, through case studies. Some (e.g., Bob McCauley 1988, Hilary Putnam 1983 and some feminist philosophers of science) argue that scientific goals cannot be understood or debated naturalistically; chapter 2 is my counterargument to their positions.

4. Most of the work on disunity (e.g., Cartwright (most recently 1999), Dupre (1993) and Rosenberg (1994)) has addressed the ontology, rather than the epistemology of science. Exceptions are Bechtel and Richardson (1993), the final chapter of Dupre (1993), and perhaps Galison (1997).

5. I discuss the generality assumption at length in my "Naturalism and Generality" (1995a).

6. I have claimed (Solomon 1995b) that this is basically Kitcher's position in *The Advancement of Science,* where I call Kitcher's position "Legend Naturalism."

7. Some experimental studies show the ways in which social interaction increases individual "bias," e.g., by anchoring effects (see, e.g., Hill 1982).

8. Goldman, Hull and Kitcher *might* say that these kinds of "hot" cognition need to be tested or analyzed in the way that economic incentives have been. The question is, why haven't they been? Why has it made sense to look at economic incentives first?

9. This is *not* the traditional internal/external distinction, which is a distinction between traditional precepts of scientific methodology (internal factors) and all other influences on science (external factors). In Sapp, the view may come from Latour and Bordieu (see Sapp 1987, p. 223).

10. Steven Stich (colloquium presentation, APA Eastern Division Meetings, December 1999) uses a similar framework for description of the various tasks of epistemology, which he takes from Richard Samuels. Epistemology is made up of descriptive, normative, evaluative and ameliorative projects.

11. I focus on feminist epistemologies of science, rather than feminist epistemologies generally. Thus, I will not discuss the ideas of Code, Jagger, Scheman, etc. Their ideas and insights are not always applicable to science, nor are they as committed to a naturalistic approach.

12. It is very difficult to classify feminist philosophies of science. Terms such as "standpoint feminist" and "feminist empiricist" are contested, reappropriated, and variously defined. I have tried to work with a noncontroversial use of the terms, in part by stating and using only minimal definitions of positions discussed. Complexities are handled not by more classificatory terms but simply through detailed examination of individual positions.

13. Sandra Harding has a better name for this position: "spontaneous feminist empiricism" (Harding 1993).

14. The exception, in naturalistic feminist empiricists, is bias conducive to truth, which should be encouraged rather than eliminated. See, for example, Antony 1993. Note that in practice feminist empiricists find few such biasing factors, and only find them among the cold cognitive biases.

15. At its most extreme, feminist postmodernism collapses into general postmodern relativism, which is considered in this book under the category of radical social constructivism. Feminist postmodernism falls under the category of feminist standpoint theory when there are non-relativistic claims about epistemic benefits of particular standpoints.

16. Kitcher (1993) even uses a military analogy.

17. Ronald Giere has also recently (1999) embraced a pluralistic position, which he calls "pluralistic perspectival realism." It is expressed in terms of a metaphor of multiple perspectives on reality. It is a weak form of pluralism, because it does not tolerate inconsistencies between the various theories that can be concurrently held. Rather, it sees the theories as covering different domains, or different parts of the same domain.

18. Some feminist epistemologists of science (e.g., Haraway, Harding, possibly Longino, Nelson) argue that scientific goals and political goals cannot be distinguished from one another. Here there is a substantive disagreement with social empiricism, and one that I implicitly argue against in chapters 2 and 3.

19. As philosopher of science, my interest is in developing a normative model that further scientific goals, i.e., social empiricism. As feminist, I share many of the other goals that Longino and other feminist philosophers of science mention,

such as better healthcare for women, special attention to feminist ideas in science (more than would occur under social empiricism alone), a more democratic scientific community, etc.

20. These points obviously need to be argued in more detail. Briefly, universal standards can produce better outcomes than leaving decisions up to individual physicians, even when the universal standards are not the best available or the only available standards. And in the absence of protocols established and agreed upon by physicians, there is a danger of insurance companies feeling justified in making their own decisions about proper medical practice—usually choosing the least costly of the available options. (For more detail, see Solomon 1998, manuscript).

21. See Catherine Hundleby's work (manuscript, 1999).

22. Democratization of a scientific community is not a straightforward idea, since it depends on particular conceptions of democracy. See Nancy McHugh (1999).

23. To be technical about it, analogical reasoning is driven by the representativeness heuristic, with highly fallible results.

References

Antony, Louise (1993). "Quine as Feminist: The Radical Import of Naturalized Epistemology." In Louise M. Antony and Charlotte Witt (eds.), *A Mind of One's Own: Feminist Essays on Reason and Objectivity,* pp. 185–225. Boulder: Westview Press.

Antony, Louise (1995). "Sisters, Please, I'd Rather Do It Myself: A Defense of Individualism in Feminist Epistemology." *Philosophical Topics* 23, no. 2: 59–93.

Bacon, Frances (1960 [1620]). *The New Organon.* Edited by Fulton Anderson. Indianapolis, Ind.: Bobbs-Merrill Educational Publishing.

Barnes, Barry, and Bloor, David (1982). "Relativism, Rationalism, and the Sociology of Knowledge." In Martin Hollis and Steven Lukes (eds.), *Rationality and Relativism* (1982), pp. 21–47. Cambridge: MIT Press.

Bechtel, William, and Richardson, Robert (1993). *Discovering Complexity: Decomposition and Localization as Strategies in Scientific Research.* Princeton, N.J.: Princeton University Press.

Bliss, Michael (1982). *The Discovery of Insulin.* Chicago: University of Chicago Press.

Bowler, Peter (1983). *The Eclipse of Darwinism.* Baltimore: Johns Hopkins University Press.

Bowler, Peter (1988). *The Non-Darwinian Revolution: Reinterpreting a Historical Myth.* Baltimore: Johns Hopkins University Press.

Bowler, Peter (1989). *The Mendelian Revolution: The Emergence of Hereditarian Concepts in Modern Science and Society.* Baltimore: Johns Hopkins University Press.

Boyd, Richard (1980). "Scientific Realism and Naturalistic Epistemology." In P. D. Asquith and R. N. Giere (eds.), *PSA 1980,* vol. 2, pp. 613–662. East Lansing, Michigan: Philosophy of Science Association.

Burian, Richard (1986). "Lillie's Paradox—Or, Some Hazards of Cellular Geography." Typescript.

Burian, Richard (1996). "Underappreciated Pathways toward Molecular Genetics as Illustrated by Jean Brachet's Cytochemical Embryology." In Sahotra Sarkar (ed.), *The Philosophy and History of Molecular Biology: New Perspectives*, pp. 67–85. Netherlands: Kluwer Academic Publishers.

Burian, Richard; Gayon, Jan; and Zallen, Doris (1988). "The Singular Fate of Genetics in the History of French Biology, 1900–1940." *Journal of the History of Biology* 21, no. 3.

Butterfield, Herbert (1965 [1931]). *The Whig Interpretation of History.* New York: W. W. Norton and Co.

Callon, Michel (1986). "Some Elements of a Sociology of Translation: Domestication of the Scallops and the Fishermen." In John Law (ed.), *Power, Action, and Belief: A New Sociology of Knowledge?* pp. 196–229. Sociological Review Monograph, no. 32. London: Routledge and Kegan Paul. Abridged and reprinted in Mario Biagoli (ed.), *The Science Studies Reader,* pp. 67–83. New York: Routledge, 1999.

Carozzi, Albert (1985). "The Reaction in Continental Europe to Wegener's Theory of Continental Drift." *Earth Sciences History* 4, no. 2: 122–137.

Cartwright, Nancy (1983). *How the Laws of Physics Lie.* New York: Oxford University Press.

Cartwright, Nancy (1999). *The Dappled World: A Study of the Boundaries of Science.* Cambridge: Cambridge University Press.

Collins, Harry (1992 [1985]). *Changing Order: Replication and Induction in Scientific Practice.* Chicago: University of Chicago Press.

Collins, Harry, and Pinch, Trevor (1993). *The Golem: What Everyone Should Know about Science.* Cambridge: Cambridge University Press.

Collins, Patricia Hill (1986). "Learning from the Outsider Within the Sociological Significance of Black Feminist Thought." *Social Problems* 33, no. 6: 14–32.

Cushing, James (1994). *Quantum Mechanics: Historical Contingency and the Copenhagen Hegemony.* Chicago: University of Chicago Press.

Darden, Lindley (1991). *Theory Change in Science: Strategies from Mendelian Genetics.* Oxford: Oxford University Press.

Darwin, Charles (1859). *The Origin of Species.* Penguin Books, 1968.

Dawes, Robyn (1988). *Rational Choice in an Uncertain World.* Orlando, Florida: Harcourt Brace Jovanovich.

Dupre, John (1993). *The Disorder of Things: Metaphysical Foundations of the Disunity of Science.* Cambridge: Harvard University Press.

Faust, David (1984). *The Limits of Scientific Reasoning.* Minneapolis: University of Minnesota Press.

Ferguson, John H. (1993). "NIH Consensus Conferences: Dissemination and Impact." In Kenneth S. Warren and Frederick Mosteller (eds.), *Doing More Good than Harm: The Evaluation of Health Care Interventions.* Vol. 703 of the Annals of the New York Academy of Sciences, Dec. 31, 1993, pp. 180–198.

Feyerabend, Paul (1975). *Against Method.* London: Verso.

Fine, Arthur (1986). *The Shaky Game: Einstein, Realism, and the Quantum Theory.* Chicago: University of Chicago Press.

Frankel, Henry (1979). "The Career of Continental Drift Theory: An Application of Imre Lakatos' Analysis of Scientific Growth to the Rise of Drift Theory." *Studies in the History and Philosophy of Science* 10: 21–66.

Frankel, Henry (1987). "The Continental Drift Debate." In H. Engelhardt and A. Caplan (eds.), *Scientific Controversies,* pp. 203–248. Cambridge: Cambridge University Press.

Fuller, Steve (1988). *Social Epistemology.* Bloomington and Indianapolis: Indiana University Press.

Galison, Peter (1997). *Image and Logic: A Material Culture of Microphysics.* Chicago: University of Chicago Press.

Giere, Ronald (1988). *Explaining Science: A Cognitive Approach.* Chicago: University of Chicago Press.

Giere, Ronald (1999). *Science without Laws.* Chicago: University of Chicago Press.

Gilbert, Scott, et al. (1988). "The Importance of Feminist Critique for Contemporary Cell Biology." *Hypatia* 3, no. 1. Reprinted in Nancy Tuana (ed.), *Feminism and Science,* pp. 172–187. Bloomington and Indianapolis: Indiana University Press, 1989.

Glen, William (1982). *The Road to Jaramillo: Critical Years of the Revolution in Earth Science.* Stanford: Stanford University Press.

Glymour, Clark (1980). *Theory and Evidence.* Princeton: Princeton University Press.

Goldman, Alvin (1992). *Liaisons: Philosophy Meets the Cognitive and Social Sciences.* Cambridge: MIT Press.

Goldman, Alvin (1999). *Knowledge in a Social World.* Oxford: Oxford University Press.

Gould, Stephen Jay (1977). *Ever Since Darwin: Reflections in Natural History.* New York: W. W. Norton and Co.

Gould, Stephen Jay (1980). *The Panda's Thumb: More Reflections in Natural History.* New York: W. W. Norton and Co.

Hacking, Ian (1983). *Representing and Intervening: Introductory Topics in the Philosophy of Natural Science.* Cambridge: Cambridge University Press.

Haraway, Donna (1991). *Simians, Cyborgs, and Women: The Reinvention of Nature.* New York: Routledge.

Harding, Sandra (1991). *Whose Science? Whose Knowledge? Thinking from Women's Lives.* Ithaca: Cornell University Press.

Harwood, Jonathan (1993). *Styles of Scientific Thought: The German Genetics Community 1900–1933.* Chicago: University of Chicago Press.

Hempel, Carl (1966). *Philosophy of Natural Science*. Foundations of Philosophy Series. Englewood Cliffs, N.J.: Prentice-Hall.

Hill, G. (1982). "Group vs. Individual Performance. Are *N* + 1 Heads Better than One?" *Psychological Bulletin* 91: 517–539.

Howson, Colin, and Urbach, Peter (1989). *Scientific Reasoning: The Bayesian Approach*. LaSalle, Ill.: Open Court.

Huizenga, John (1992). *Cold Fusion: The Scientific Fiasco of the Century*. Rochester: University of Rochester Press.

Hull, David (1988). *Science as a Process: An Evolutionary Account of the Social and Conceptual Development of Science*. Chicago: University of Chicago Press.

Hume, David (1977 [1748]). *An Enquiry Concerning Human Understanding*. Indianapolis, Ind.: Hackett Publishing Co.

Hundleby, Catherine (1999). "A Naturalist History of Feminist Standpoint Theory." Manuscript.

Kahneman, Daniel; Slovic, Paul; and Tversky, Amos (eds.), (1982). *Judgments under Uncertainty: Heuristics and Biases*. Cambridge: Cambridge University Press.

Keller, Evelyn Fox (1983). *A Feeling for the Organism: The Life and Work of Barbara McClintock*. New York: W. H. Freeman and Co.

Keller, Evelyn Fox (1985). *Reflections on Gender and Science*. New Haven and London: Yale University Press.

Kevles, Daniel (1985). *In the Name of Eugenics: Genetics and the Uses of Human Heredity*. New York: Alfred A. Knopf.

Kevles, Daniel (1995). "Pursuing the Unpopular: A History of Courage, Viruses, and Cancer." In Robert Silvers (ed.), *Hidden Histories of Science*, pp. 69–112. New York: New York Review of Books.

Kitcher, Philip (1985). *Vaulting Ambition: Sociobiology and the Quest for Human Nature*. Cambridge: MIT Press.

Kitcher, Philip (1990). "The Division of Cognitive Labor." *Journal of Philosophy* 87, no. 1: 5–22.

Kitcher, Philip (1993). *The Advancement of Science*. Oxford and New York: Oxford University Press.

Kohler, Robert (1994). *Lords of the Fly: Drosophila Genetics and the Experimental Life*. Chicago: University of Chicago Press.

Kornblith, Hilary (1985). "Introduction: What Is Naturalistic Epistemology?" In H. Kornblith, *Naturalizing Epistemology*, pp. 1–14. Cambridge, Mass: MIT Press. Second ed.: 1994.

Kuhn, Thomas (1970 [1962]). *The Structure of Scientific Revolutions*. Second ed. Chicago: University of Chicago Press.

Kuhn, Thomas (1977). "Objectivity, Value Judgment, and Theory Choice." In Thomas Kuhn, *The Essential Tension*, pp. 320–339. Chicago: University of Chicago Press.

Lakatos, Imre (1970). "Falsification and the Methodology of Scientific Research Programmes." In Imre Lakatos and Alan Musgrave (eds.), *Criticism and the Growth of Knowledge,* pp. 91–196. Cambridge: Cambridge University Press.

Latour, Bruno (1987). *Science in Action.* Cambridge: Harvard University Press.

Latour, Bruno (1988). *The Pasteurization of France.* Cambridge: Harvard University Press.

Laudan, Larry (1977). *Progress and Its Problems: Towards a Theory of Scientific Growth.* Berkeley and Los Angeles: University of California Press.

Laudan, Larry (1981). "A Confutation of Convergent Realism." *Philosophy of Science* 48: 19–48. Reprinted in R. Boyd, P. Gasper, and J. D. Trout (eds.), *The Philosophy of Science,* pp. 223–245. Cambridge: MIT Press, 1991.

Laudan, Larry (1984). *Science and Values: The Aims of Science and Their Role in Scientific Debate.* Berkeley and Los Angeles: University of California Press.

Laudan, Rachel (1978). "The Recent Revolution in Geology and Kuhn's Theory of Scientific Change." In P. D. Asquith and I. Hacking (eds.), *PSA 1978,* vol. 2, pp. 227–239. East Lansing: Philosophy of Science Association.

Laudan, Rachel, and Laudan, Larry (1989). "Dominance and the Disunity of Method: Solving the Problems of Innovation and Consensus." *Philosophy of Science* 56, no. 2: 221–237.

Le Grand, H. E. (1988). *Drifting Continents and Shifting Theories.* Cambridge: Cambridge University Press.

Lillie, Frank (1927). "The Gene and the Ontogenic Process." *Science* 64: 361–368.

Lloyd, Elisabeth (1997). "Feyerabend, Mill, and Pluralism." *Philosophy of Science 64* (*Proceedings of PSA*), pp. S396–S407.

Longino, Helen (1990). *Science as Social Knowledge: Values and Objectivity in Scientific Inquiry.* Princeton: Princeton University Press.

Longino, Helen (1994). "In Search of Feminist Epistemology." *Monist* 77, no. 4: 472–485.

Longino, Helen (1994). "The Fate of Knowledge in Social Theories of Science." In Frederick Schmitt (ed.), *Socializing Epistemology: The Social Dimensions of Knowledge,* pp. 135–157. Lanham, Md.: Rowman and Littlefield.

Marsh, Margaret, and Ronner, Wanda (1996). *The Empty Cradle: Infertility in America from Colonial Times to the Present.* Baltimore: Johns Hopkins University Press.

McCauley, Robert (1988). "Epistemology in an Age of Cognitive Science." *Philosophical Psychology* 1, no. 2: 143–152.

McEvoy, John (1983). "Enlightenment and Dissent in Science: Joseph Priestley and the Limits of Theoretical Reasoning." *Enlightenment and Dissent* 2: 47–67.

McHugh, Nancy (1999). "Toward a More Democratic Science." Doctoral dissertation, Temple University.

McMullin, Ernan (1987). "Scientific Controversy and Its Termination." In H. Tristram Engelhardt and Arthur Caplan (eds.), *Scientific Controversies: Case Studies in the Resolution and Closure of Disputes in Science and Technology*, pp. 49–91. Cambridge: Cambridge University Press.

Menard, H. W. (1986). *The Ocean of Truth: A Personal History of Global Tectonics*. Princeton: Princeton University Press.

Mill, John Stuart (1859). *On Liberty*. Reprinted in Mary Warnock (ed.), *John Stuart Mill: Utilitarianism* (1962). Glasgow, UK: William Collins Sons and Co.

Nelson, Lynn Hankinson (1990). *Who Knows: From Quine to a Feminist Empiricism*. Philadelphia: Temple University Press.

Nersessian, Nancy (1992). "How Do Scientists Think? Capturing the Dynamics of Conceptual Change in Science." In Ronald Giere (ed.), *Cognitive Models of Science*, pp. 3–44. Minneapolis: University of Minnesota Press.

Nickles, Thomas (1992). "Good Science as Bad History: From Order of Knowing to Order of Being." In Ernan McMullin (ed.), *The Social Dimensions of Science*, pp. 85–129. Notre Dame: University of Notre Dame Press.

Nisbett, Richard, and Ross, Lee (1980). *Human Inference: Strategies and Shortcomings of Social Judgment*. Englewood Cliffs: Prentice-Hall.

Nisbett, Richard, et al. (forthcoming). "Culture and System of Thoughts: Holistic versus Analytic Cognition." *Psychological Review*.

Odroyd, David (1984). "How Did Darwin Arrive at His Theory?" *History of Science* 22: 325–374.

Okin, Susan (1989). *Justice, Gender and the Family*. Basic Books.

Olby, Robert (1974). *The Path to the Double Helix*. Seattle: University of Washington Press.

Oreskes, Naomi (1999). *The Rejection of Continental Drift: Theory and Method in American Earth Science*. New York: Oxford University Press.

Pickering, A. (1995). *The Mangle of Practice: Time, Agency, and Science*. Chicago: University of Chicago Press.

Profet, Margie (1993). "Menstruation as a Defense against Pathogens Transported by Sperm." *Quarterly Review of Biology* 68: 5–71.

Psillos, Stathis (1999). *Scientific Realism: How Science Tracks Truth*. London and New York: Routledge Press.

Putnam, Hilary (1975). *Mathematics, Matter and Method*. Vol. 1 of *Philosophical Papers*. Cambridge: Cambridge University Press.

Putnam, Hilary (1983). *Realism and Reason*. Vol. 3 of *Philosophical Papers*. Cambridge: Cambridge University Press.

Quine, W. V. (1960). *Word and Object*. Cambridge: MIT Press.

Quine, W. V. (1969). "Epistemology Naturalized." In his *Ontological Relativity and Other Essays*. New York: Columbia University Press.

Reichenbach, Hans (1938). *Experience and Prediction.* Chicago: University of Chicago Press.

Rosenberg, Alexander (1994). *Instrumental Biology and the Disunity of Science.* Chicago: University of Chicago Press.

Rosenberg, Charles (1976). *No Other Gods: On Science and American Social Thought.* Baltimore: Johns Hopkins University Press.

Sandel, Michael (1982). *Liberalism and the Limits of Justice.* Cambridge: Cambridge University Press. Second ed., 1998.

Sapp, Jan (1987). *Beyond the Gene: Cytoplasmic Inheritance and the Struggle for Authority in Genetics.* New York and Oxford: Oxford University Press.

Sarkar, Husain (1983). *A Theory of Method.* Berkeley and Los Angeles: University of California Press.

Shapin, Steven, and Schaffer, Simon (1985). *Leviathan and the Air-Pump: Hobbes, Boyle, and the Experimental Life.* Princeton: Princeton University Press.

Shields, Stephanie (1982). "The Variability Hypothesis: The History of a Biological Model of Sex Differences in Intelligence." Reprinted in Sandra Harding and Jean O'Barr (eds.), *Sex and Scientific Inquiry,* pp. 187–215. Chicago: University of Chicago Press, 1987.

Solomon, Miriam (1992). "Scientific Rationality and Human Reasoning." *Philosophy of Science* 59, no. 3: 439–455.

Solomon, Miriam (1993). "Is There an Invisible Hand of Reason?" Manuscript written for the international conference "Non-formal Foundations of Reason," held in Newcastle, Australia, August 1993.

Solomon, Miriam (1994a). "Social Empiricism." *Noûs* 27, no. 3: 325–343.

Solomon, Miriam (1994b). "A More Social Epistemology." In Fred Schmitt (ed.), *Socializing Epistemology: The Social Dimensions of Knowledge,* pp. 217–233. Lanham, Md.: Roman and Littlefield.

Solomon, Miriam (1995a). "Naturalism and Generality." *Philosophical Psychology* 8, no. 4: 353–363.

Solomon, Miriam (1995b). "Legend Naturalism and Scientific Progress." *Studies in History and Philosophy of Science* 26, no. 2: 205–218.

Solomon, Miriam (1995c). "The Pragmatic Turn in Naturalistic Philosophy of Science." *Perspectives on Science* 3, no. 2: 206–230.

Solomon, Miriam (1996). "Social Epistemology." In Donald M. Borchert (ed.), *Encyclopedia of Philosophy, Supplement,* pp. 546–548. New York: Simon and Schuster, Macmillan.

Solomon, Miriam (1998). "The Medical Consensus Conference." Manuscript for discussion at the Philosophy of Science Association Meetings, October 1998.

Sonneborn, Tracy (1950). "Partner of the Genes." *Scientific American,* November 1950, pp. 30–39.

Stewart, Jay (1990). *Drifting Continents and Colliding Paradigms.* Bloomington and Indianapolis: Indiana University Press.

Stich, Stephen (1990). *The Fragmentation of Reason.* Cambridge: MIT Press.

Sulloway, Frank (1996). *Born to Rebel: Birth Order, Family Dynamics and Creative Lives.* New York: Pantheon Books.

Taubes, Gary (1993). *Bad Science: The Short Life and Weird Times of Cold Fusion.* New York: Random House.

Thagard, Paul (1989). "Scientific Cognition: Hot or Cold?" In Steve Fuller et al. (eds.), *The Cognitive Turn: Sociological and Psychological Perspectives on Science,* pp. 71–82. Dordrecht, Netherlands: Kluwer Academic Publishers.

Thagard, Paul (1992). *Conceptual Revolutions.* Princeton: Princeton University Press.

Thagard, Paul (1993). "Societies of Minds: Science as Distributed Computing." *Studies in the History and Philosophy of Science* 24, no. 1: 49–67.

Van Fraassen, Bas (1980). *The Scientific Image.* New York: Oxford University Press.

Vertosick, Frank, Jr. (1998). "First, Do No Harm." In *Discover,* July 1998, pp. 106–111.

Wegener, Alfred (1966 [1915]). *The Origin of Continents and Oceans.* Fourth ed. Translated by J. Biram. London: Dover.

Woolgar, Steve (1988). *Science: The Very Idea.* Chichester, England: Ellis Horwood.

Wylie, Alison (2000). "Rethinking Unity as a 'Working Hypothesis' for Philosophy of Science: How Archaeologists Exploit the Disunities of Science." *Perspectives on Science* 7, no. 3: 293–317.

Wylie, Alison, and Nelson, Lynn (1998). "Coming to Terms with the Values of Science." Talk at the Center for Philosophy of Science, University of Pittsburgh.

Young, Iris (1990). *Justice and the Politics of Difference.* Princeton: Princeton University Press.

Young, Robert (1995). *Darwin's Metaphor: Nature's Place in Victorian Culture.* Cambridge: Cambridge University Press.

Index

Printed in the United States
By Bookmasters